SCIENCE

100 SCIENTISTS WHO CHANGED THE WORLD

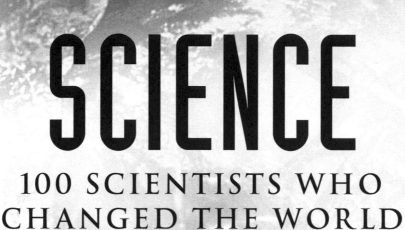

SCIENCE

100 SCIENTISTS WHO CHANGED THE WORLD

JON BALCHIN

ARCTURUS

CONTENTS

Arcturus Publishing Ltd
26/27 Bickels Yard
151–153 Bermondsey Street
London SE1 3HA

Published in association with
foulsham
W. Foulsham & Co. Ltd,
The Publishing House, Bennetts Close, Cippenham,
Slough, Berkshire SL1 5AP, England

ISBN 0-572-02934-9

Editor: Paul Whittle; Design: Alex Ingr

Acknowledgements
The publishers would like to thank Glen Carlstrom and Virginia Ingr for
their invaluable contributions to this book.

The author would like to thank the following people: Anne Fennell, Paul
Whittle, Matthew Smith and everyone at Arcturus Publishing, KTB, my
parents, Jon, Iain, Faye, Grace, Alice, Alan, Irene, my extended family
across the world, and last but never least Merryn.

Printed in China

FOREWORD

To be alive today is to be confronted by the products of science. Science has given us television, the internal combustion engine, the aeroplane, and the computer, to name but a few. Yet consumer products such as these are but one aspect of the benefits science can bring to mankind. Too often, for example, the field of medicine is overlooked in favour of more 'glamorous' fields, such as astrophysics or rocketry.

As recently as the last century, death from disease was an everyday occurrence. Both small-pox and polio killed millions until Edward Jenner made the simple yet life-changing discovery that milkmaids infected with cowpox were immune from smallpox, and Jonas Salk developed the polio vaccine. That both diseases continue to be killers in the modern world is due not to science, but to a tragic reluctance on the part of richer countries to share its benefits with their poorer counterparts.

Science has also produced less beneficial developments; the tank, machine gun, and atomic bomb, but science does achieve results, however morally questionable some of those results may be, and it is this which sets it apart from superstition, witchcraft and religion.

Important though the products of science may be, what is perhaps more significant is the scientific method itself, proceeding as it does from empirical observation to theory, to modification of theory in the light of further evidence.

We may still pray for rain, but we understand the physical causes of the weather, and to an extent can predict it; we no longer ascribe it to the actions of some unknowable deity and sacrifice our first-born in the hope of a favourable outcome.

This method contrasts with the previous means of discovering truth 'by authority', which claimed beliefs as true not on the basis of what was claimed, rather on the basis of who was making the claim.

Rejecting the notion of truth by authority, the scientists in this book observed the world around them, proposed theories to explain it, and modified these theories to account for further observations.

The road out from the darkness of superstition into the light of reason has not always been an easy one. When Vesalius dared to contradict the authority of Galen, he was abused as a liar and madman; the Montgolfier brothers' claims met only scepticism. Galileo and Copernicus both narrowly avoided following Giordano Bruno to the stake for proposing the heliocentric theory of the solar system, in opposition to accepted Church dogma. Yet they perservered, and in so doing, lit a beacon for the rest of humanity to follow.

The men and women who make up this book have blazed, in Bertrand Russell's poetic phrase, 'with all the noonday brightness of human genius'. How far the beacon they have lit will guide us, and how far science will yet progress, however, we shall leave to the next generation of scientists who will change the world.

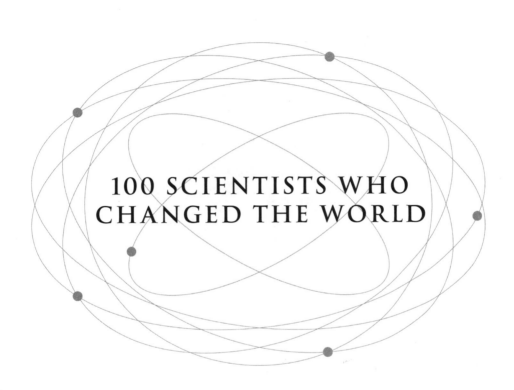

100 SCIENTISTS WHO
CHANGED THE WORLD

ANAXIMANDER

C. 611–547BC

A NOTE ON DATES   Beyond the fact that Anaximander was born in the Greek city of Miletus, on the coast of Asian Turkey, probably around 611 BC, we know very little of his life. This is mainly because he wrote very little down, a task he preferred to leave to his pupils. What we do know has come down to us secondhand through later Greek scientist-philosophers, who naturally took an interest in their illustrious predecessors.

Imagine a world everyone knew to be flat, supported in the vastness of space by pillars. It was widely accepted that this world sat at the centre of a tent-like universe, with stars, equidistant from the earth, stuck around the edges. Now imagine being informed that, contrary to popular opinion, the world had 'depth', it was completely unsupported, and the stars, moon and sun were not only at different distances away, but also cycled around this three dimensional earth. It would be a revolutionary concept, completely

altering existing preconceptions of the universe, and is exactly the giant leap in scientific understanding with which Anaximander is credited.

▶ A THEORY OF THE INFINITE

Often known as the founder of modern astronomy, Anaximander is the beginning point of the current Western concept of the universe. A Greek, he was born and died in Miletus, now in modern Turkey, although he is also thought to have travelled widely as he formulated his views

Anaximander effectively discovered the idea of space: that is, a universe with depth

of the cosmos. Anaximander was a pupil of Thales of Miletus, himself credited with original work in physics, philosophy, geometry and astronomy. As with Thales, very little detail is known about Anaximander's life and only one passage of his original texts remains. The rest of what we know comes from descriptions by later Greeks, in particular **Aristotle** and Theophrastos. If anything, they remembered Anaximander to be a philosopher rather than a scientist, expressing a bold theory on the 'infinite' or 'boundless'. This idea was his 'first principle' of all things, with no origin and no end but 'from which came all the heavens and the worlds in them' (from Theophrastos's description of Anaximander's work). Yet it is his ideas in astronomy which had the long-term impact, presenting theories which changed the world.

▶ **A TOPOGRAPHICAL UNIVERSE**

Arguably Anaximander's most important leap was to conceptualise the earth as suspended completely unsupported at the centre of the universe. It had been assumed by other Greek thinkers that the earth was a flat disc held in place by water, pillars, or another physical structure. Although Anaximander obviously had no notion of gravity, he supported his argument by supposing that the earth, being at the centre of the universe, at 'equal distances from the extremes, has no inclination whatsoever to move

up rather than down or sideways; and since it is impossible to move in opposite directions at the same time, it necessarily stays where it is' (Aristotle explaining Anaximander's theory). Moreover, because the earth was suspended freely, it allowed Anaximander to propose the idea that the sun, moon and stars orbited in full circles around the earth. This explained, for example, why the sun would disappear in the west and rise in the east. When you add to this the idea that the earth had depth (although Anaximander envisaged it being cylindrical in shape and still with a flat disc on the top which was the only 'surface') a revolutionary view of the universe emerges.

▶ **THE VOID BETWEEN THE STARS**

Anaximander effectively discovered the idea of 'space' or a universe with depth. Rather than view the earth caged in a planetarium style 'celestial vault', he argued the 'celestial bodies' (the sun, moon and stars) were different distances away from the earth, with space or air between them. He attempted to ascribe distances for these bodies from the earth as they rotated around it, although he wrongly proposed the stars were closest, then the moon, with the sun furthest away. Anaximander may have drawn a map of his version of the universe. Although wrong in its detail, this would have been a steep change in its graphical representation.

FURTHER ACHIEVEMENTS

Anaximander was not just an astronomer. He is thought to have introduced the sundial to Greece from Babylon, using it to determine the solstices and equinoxes. In geography, he is thought to have drawn the first map of the known world, a hugely important breakthrough in itself. Meanwhile, in biology, he may have pre-empted Darwin's theories of evolution, albeit unwittingly, through his belief that mankind grew out of the original animal inhabitants of the earth. Anaximander believed these to be primitive kinds of fish taking their form from rising water heated by the sun.

PYTHAGORAS

C. 581–497 BC

CHRONOLOGY • **c. 525 BC** Pythagoras is taken prisoner by the Babylonians • **c. 518 BC** establishes his own academy at Croton, now Crotone, in southern Italy, where he is regarded by many as a cult leader • **c. 500 BC** as Croton becomes increasingly politically unstable, Pythagoras makes a final move to Metapontum

Very little is known for certain about the life of this Greek mathematician and philosopher. One obstacle is that many of the mathematical discoveries credited to Pythagoras may have actually been discovered by his disciples, the Pythagoreans, members of the semi-religious, philosophical school he founded. Moreover, because of the reverence with which the originator of the brotherhood was treated by his followers and biographers, it is sometimes difficult to discern legend from fact.

▶ **EXPERIMENTAL MATHEMATICS**

It is, however, fairly clear that Pythagoras himself did undertake practical experiments concerning the relationship between mathematics and music. It is thought he either attached different weights to a series of strings, or alternatively experimented with different string lengths, examining the mathematical relationship between the resultant notes when plucked and the weights or lengths applied. What he discovered was that simple, whole number relations, for example a string of one length and another of twice that length, produced harmonious tones. These obser-

'All physical things the stars and the universe, are mathematically related'

vations ultimately led to the determination of musical scales as we know them today. Not only was this a momentous musical discovery, but was probably the first time a physical law had been mathematically expressed. As a result it began the science of mathematical physics.

▸ THE WORLD AS A SPHERE

This idea of a harmonious relationship between physical entities also enabled Pythagoras to conceptualise the world as a sphere, even if he had limited scientific basis at that time with which to back up his belief. For Pythagoras and his followers the idea of a 'perfect' mathematical interrelation between a globe moving in circles and the stars behaving similarly in a spherical universe (just as musical tones harmoniously danced around and depended on each other) seemed much more pleasing than **Anaximander's** cylindrical earth, or one composed of a flat disc. The view was so powerful that it inspired later Greek scholars, including **Aristotle**, to seek and ultimately find physical and mathematical evidence to reinforce the theory of the world as an orb.

▸ PYTHAGORAS AND HIS SCHOOL

Pythagoras founded his school at Croton in Italy, one of its objectives being to further explore the relationship between the physical world and mathematics. Indeed, of the five key beliefs that the Pythagoreans held, one was dominant: the idea that 'all is number'. In other words, reality is at its fundamental level mathematical and that all physical things, like musical scales or the spherical earth and its companions the stars and the universe, are mathematically related. The experiments of the Pythagoreans led to numerous discoveries such as 'the sum of a triangle's angles is the equal to two right angles (180°)'. 'The sum of the interior angles in a polygon of n sides is equal to 2n-4 right angles' was another. Yet arguably their most important arithmetical discovery was that of irrational numbers. This came from the realisation that the square root of two could not be expressed as a perfect fraction. This was a major blow to the Pythagorean idea of perfection and according to some accounts attempts were even made to try to conceal the discovery.

▸ PYTHAGORAS' THEOREM

Pythagoras' famous Theorem was probably known to the Babylonians but Pythagoras may well have been the first to mathematically prove it. 'The square of the hypotenuse on a right-angled triangle is equal to the sum of the squares on the other two sides' can otherwise be expressed as $a^2+b^2=c^2$, where a and b are the shorter sides of the triangle and c is the hypoteneuse.

FURTHER ACHIEVEMENTS

It is perhaps ironic that Pythagoras is remembered today for his Theorem, the principles of which had previously been known for over a thousand years, and yet his more original discoveries are obscure. As the discoverer of the musical scale, in effect creating a rule book for the musical harmonies that we take for granted, *it is arguable that this has had a much more profound impact on the history of the world than a simple, largely borrowed, mathematical formula. Equally, some 2,000 years before Christopher Columbus was credited with the idea, Pythagoras proposed that the world was a sphere.*

HIPPOCRATES OF COS

C. 460–377BC

A NOTE ON DATES   Beyond the fact that Hippocrates was born on Cos, probably around the middle of the fifth century BC, such dates as we have, as for Anaximander, are generally so vague as to be scarcely worth mentioning.

Much of what is attributed to Hippocrates is contained within *The Hippocratic Collection*, a series of sixty to seventy medical texts written in the late fifth and early fourth century BC. It is widely acknowledged, though, that Hippocrates himself could not have written many of these works, which is why precise details about his life and achievements remain unclear. Written over the period of a century and varying hugely in style and argument, it is thought they came from the medical school library of Cos, possibly put together in the first instance by the author to whom they later became attributed.

Given the sobriquet of the 'Great Physician' by Aristotle, Hippocrates is today more commonly referred to as the 'Father of Medicine'. Without question, Hippocrates of Cos, in spite of the limited factual details actually known about his life, helped lay the foundation stones of the science of medicine and greatly influenced its later development, even up to the present day.

▸ A COMMON SENSE APPROACH

For Hippocrates, disease and its treatment were entirely of this earth. He cast aside superstition and focused on the natural, in particular observ-

The answers he prescribed are still good medicine two thousand years later

ing, recording and analysing the symptoms and passages of diseases. The prognosis of an illness was central to Hippocrates's approach to medicine, partly with a view to being able to avoid in future the circumstances which were perceived to have initiated the problems in the first place. The development of far-fetched cures or drugs, however, was not so important. What came from nature should, in Hippocrates's mind, be cured by nature; therefore rest, healthy diet, exercise, hygiene and air were prescribed for the treatment and prevention of illness. 'Walking is a man's best medicine,' Hippocrates wrote.

▶ THE THEORY OF HUMOURS

He regarded the body as a single entity, or whole, and the key to remaining healthy lay in preserving the natural balance within this entity. The four 'humours' he believed influenced this equilibrium were blood, phlegm and yellow and black bile. When present in equal quantities, a healthy body would result. If one element became too dominant, however, then illness or disease would take over. The way to treat the problem would be by trying to undertake activities or eat foods which would stimulate the other humours, while at the same time attempting to restrain the dominant one, in order to restore the balance and, consequently, health.

Although this approach may still seem a little unscientific by today's standards of medicine, the fact that Hippocrates was prescribing such a natural, 'earthly' solution at all was a major advancement. Moreover, the concept and treatment of humours endured for the next two thousand years, certainly as far as the seventeenth century and in some aspects as far as the nineteenth. In addition, the answers he prescribed for healthy living such as diet and exercise are still 'good medicine' two thousand years later. Language introduced by Hippocrates also endures: an excess of black bile in Greek was 'melancholic', while someone with a too dominant phlegm humour became 'phlegmatic'.

▶ THE OATH OF HIPPOCRATES

Ironically, Hippocrates may not have even written his own most enduring legacy. The Hippocratic Oath, probably penned by one of his followers, is a short passage constituting a code of conduct to which all physicians were henceforth obliged to pledge themselves. It outlines, amongst other things, the ethical responsibilities of the doctor to his patients and a commitment to patient confidentiality. It was an attempt to set physicians in the Hippocratic tradition apart from the spiritual and superstitious healers of their day. Such has been its durability that even students graduating from medical school today can still vow the Oath.

THE LEGACY OF HIPPOCRATES

Before the time of Hippocrates there had been virtually no science at all in medicine. Disease was believed to be the punishment of the gods, Divine intervention came not from the natural, but the supernatural. The only 'treatment', therefore, also came from the supernatural: through magic, witchcraft, superstition or religious ritual.

Hippocrates confronted this notion head on, with a conviction remarkable given the age in which he lived. His approach brought the rational to the previously irrational and with it medicine strode into the age of reason. 'There are in fact two things,' said Hippocartes of Cos, 'science and opinion; the former begets knowledge, the latter ignorance.'

DEMOCRITUS OF ABDERA

c. 460–370BC

John **Dalton** is widely remembered today as the founder of atomic theory for his work in the nineteenth century which proposed that elements were made of tiny, indivisible particles. Yet the concept of the 'atom', and a systematic argument for how it formed the physical world had been in existence for over two thousand years before him, expounded by Democritus of Abdera, in Thrace.

▶ **ATOMIC THEORY**

The word 'atom' comes from the Greek *atomon*, meaning 'indivisible'. Dalton acknowledged this two millennia later by using the same word in his thesis. But even Democritus was not the first. His teacher, Leucippus, as well as Anaxagoras, had all considered this notion of the indivisible particle.

Democritus, however, was the first to propose an all-encompassing argument for the primacy of the atom in the make up of the universe. Although not based on scientific evidence, as was Dalton's, instead being simply a reasoned hypothesis, many aspects of Democritus's theories are still resonant.

Democritus presents a systematic argument for the primacy of the atom

▶ ATOMS, BEING AND THE VOID

For Democritus, there were only two things: space and atoms. Space consisted of the 'Void', an infinitely sized vacuum, with infinite numbers of atoms making up 'Being', or the physical world. The atoms and space had both always existed and always would exist because nothing could come from nothing. The atoms, which were the constituent parts of everything on earth, as well as of the planets and the stars, had always and would always be the same: solid, impenetrable, invisible blocks which never changed. They simply combined with other atoms in the Void to form different things from rocks to plants to animals. When these things died or fell apart the structure disintegrated and the atoms were free to form new things by combining again in a different shape with other atoms.

Democritus reasoned that the method by which the atoms could combine was through their different shapes. While all atoms were the same in substance, liquids were thought to have smooth, round edges so they could fall over each other, while those which made up solids had toothed, rough edges which could hook on to each other. As with physical form, Democritus argued that other perceived differences in things, such as their taste, could also be explained by the edges of the atoms: sweet tastes were caused by large round atoms, sharp ones by jagged, heavy atoms. Likewise, the colour of things was explained by the 'position' of the atoms within a compound which would result in darker or lighter shades or shadows being cast.

Democritus's thesis is all the more remarkable because it completely rejects the notion of the spiritual or religious. The soul, for example, was explainable through a fast-moving group of atoms brought together by encasement in the body. These atoms reacted to disturbances by other atoms inside and outside that body. The motion produced sensations which interacted with the mind (itself just a collection of atoms) to produce thoughts, feelings and so on. Once the body was dead, however, Democritus argued that the soul ceased to exist because the object which held the fast-moving atoms together had disintegrated. Thus released, they could separate and interact with other atoms to form new things. This left no place for abstract notions of the supernatural or an afterlife.

▶ DETERMINISM

Even the concept of freedom of choice could not exist under Democritus's model. All human actions were determined by atoms striking the human body, not as part of any grand design or plan, but simply because motion and collision with other atoms within the Void is what they did and always would do, thereby leaving no free will to the human at all.

A MATHEMATICAL LEGACY

Although many elements of Democritus's thesis have subsequently been tested and often discredited by modern science, his remains one of the earliest attempts to explain the universe with a few simple physical and mathematical laws. This represents an important change in thinking towards the subject and is a notion that has pre-occupied scientists ever since.

Democritus is also credited with discovering the mathematical law that a cone's volume is a third that of a cylinder sharing the same sized base and height, as well as a similar respective relationship between a pyramid and a prism.

15

PLATO

C. 427–347 BC

CHRONOLOGY • **427** BC Plato born in or around Athens • **399** BC On the execution of Socrates, Plato leaves Athens in disgust • **387** BC On his return to Athens, Plato founds his Academy, a bastion of intellectual achievement until its closure on the orders of the emperor Justinian in 529 AD • **389** BC Plato visits Sicily for the first time

To understand how Plato came to the conclusions which have exercised such a profound impact on Western thinking, it is necessary to understand his own influences. Born in or around Athens at a time when the city-state was flourishing as one of the most dominant and culturally enlightened places on earth, he was strongly affected by the arguments of another great philosopher, Socrates, who also lived there. Socrates' approach was to constantly strive for clearer definitions of words and people's perception of those words in order to get nearer to 'the truth' that lay behind their often irrational and ill-thought-out use of them. This introduced to Plato the notion of 'reality' being distorted by human perceptions, which would become important in his approach to science and, in particular, metaphysics.

▸ SOCRATES' INFLUENCE

Socrates' was executed in 399 BC for allegedly 'corrupting' the youth of Athens with his 'rebellious' ideas. Reacting to this, Plato fled the city-state and began a tour of many countries which would last more than a decade. On his

'Let no one ignorant of geometry enter here'

INSCRIPTION ABOVE THE ENTRANCE TO PLATO'S ACADEMY

travels he encountered a group of people who would become another major influence, the Pythagoreans. Begun by their founder **Pythagoras,** the school of disciples in Croton continued to promote their 'all is number' approach to everything.

▶ THE THEORY OF FORMS

The combination of these two major forces on Plato plus, of course, his own work, brought him to his Theory of Forms, his main legacy to scientific thought. This consisted of an argument that nature, as seen through human eyes, was merely a flawed version of true 'reality' or 'forms'; in an instructive metaphor, he compares humanity with cave dwellers, who live facing the back wall of the cave. What they perceive as reality, is merely the shadows thrown out by the sun. There is, therefore, little to be learnt from direct observation of them. For Plato, there had always existed an eternal, underlying mathematical form and order to the universe, and what humans saw were merely imperfect glimpses of it, usually corrupted by their own irrational perceptions and prejudices about the way things 'are'. Consequently, for Plato, like the Pythagoreans, the only valid approach to science was a rational, mathematical one which sought to establish universal truths irrespective of the human condition. This validation of the numerical method

strongly impacted on modern science; disciples following in its tradition 'made' discoveries by mathematical prediction. For example, arithmetic calculations would suggest that future discoveries would have particular properties, in the case of unknown elements in Dmitry **Mendeleev's** first periodic table for instance, and subsequent investigative work by scientists would prove the mathematics to be true. It is an approach still used by scientists today.

▶ THE ACADEMY

Plato also helped to influence scientific thought in a much more physical sense by founding an Academy on his return to Athens in 387 BC. Some commentators claim this institute to be the first European university, and certainly its founding principles as a school for the systematic search for scientific and philosophical knowledge were consistent with such an establishment. Plato's influence was pervasive; it is said there was an inscription over the entrance to the institute which read, 'Let no one enter here who is ignorant of geometry.' Over the subsequent centuries, the Athenian Academy became recognised as the leading authority in mathematics, astronomy, science and philosophy, amongst other subjects. It survived for nearly a thousand years until the Roman emperor Justinian shut it down in 529 AD, around the time the Dark Ages began.

THE LEGACY OF PLATO

Plato is best remembered today as one of the greatest philosophers of the Western tradition. He might not, therefore, be an obvious candidate for inclusion in a book on famous scientists. But in exactly the same way that the influence of Plato's work stretched into many other academic areas such as education, literature, political

thought, epistemology and aesthetics, so it is the case with his science.

Although Plato's scientific and philosophical legacy has undergone significant revival and reinterpretation over the course of history, his logical approach to science remains influential, standing testament to his far-reaching ideas.

ARISTOTLE

C. 384–322BC

CHRONOLOGY • **367** BC Aristotle enters Plato's Academy in Athens • **347** BC On Plato's death, he leaves the Academy for Lesbos • **342** BC Becomes tutor to the young Alexander the Great at Macedon • **335** BC Returns to Athens and founds his own school, the Lyceum • **323**BC Accused of impiety: to 'prevent the city sinning twice against philosophy', Aristotle returns to Chalcis, where he dies the following year

Aristotle's work in physics and cosmology dominated Western thought until the time of **Galileo** and **Newton**, when much of it was subsequently proved to be wrong. He began with the accepted Greek notion that everything was made up of one of four elements: earth, water, air or fire.

▸ THE FOUR ELEMENTS

He also accepted the notion of the earth at the centre of the universe, with the moon, planets, sun and stars all orbiting around it in perfect circles. He believed that the four elements always sought to return to their 'natural place'. This was why a rock, for example, would drop to the earth as soon as any obstacles preventing it from doing so were removed – because 'earth' elements, being denser and heavier, would naturally seek to move downwards towards the centre of the planet. Water elements would float around the surface, air would rise above that and fire would seek to rise above them all, explaining the leaping, upward direction of flames. By the same method,

Aristotle's scientific proposals have at times been accorded an almost god-like authority

Aristotle could explain why a rock would travel through the air first before heading downwards when thrown, rather than straight towards the earth, as one would expect. This was because the air, seeking to close the gap made by the invasion of the rock, would propel it along until it lost its horizontal speed and it tumbled to the ground.

▶ THE FIFTH ELEMENT

Aristotle encountered a problem, however. This notion of everything tending towards its 'natural place' was not consistent with his view of the rest of the cosmos which was rotating in perfect, uniform order, with none of the disturbances or jostling for position associated with earthly elements (otherwise the planets and stars would tumble towards earth at the centre of the universe). To explain this, he added a fifth element to the traditional four, that of 'aether' which naturally had a circular motion. Everything beyond the moon was regulated by aether, explaining both its perfect movement and stability, while everything below it was subject to the laws of the four other elements. Although this explanation may seem far-fetched to a modern audience, it was widely accepted for the next two thousand years. In so doing it made a lasting impact on the development of scientific thought, if only in slowing down its progress due to the unchallenging acceptance with which Aristotle's laws were accepted for so long.

In other physical areas Aristotle was more accurate in his assessment. For example, he reinforced the view initially espoused by **Pythagoras** that the earth was spherical. He noticed every time there was a lunar eclipse, an arc-shaped shadow consistent with a globe was cast upon the moon. In addition, he noted correctly that when travelling north or south along the earth, stars 'moved' on the horizon until some gradually disappeared from view. He concluded that this would also be consistent with the idea of a spherical planet.

▶ TOWARDS BIOLOGY

Some of Aristotle's biology was faulty, such as the notion of the heart, not the brain, as the seat of the mind. However, consistent with his empirical approach he undertook detailed dissections to dispel certain myths, for example, that an embryo is formed at the moment of fertilisation, and that the sex of an animal is determined by its position in the womb.

Aristotle was also one of the first to attempt a methodical classification of animals, using means of reproduction, differentiating between those animals which gave birth to live young, and those which laid eggs, a system which was the forerunner of modern taxonomy.

THE LEGACY OF ARISTOTLE

In contrast to his teacher and mentor Plato, Aristotle believed there was much to be learnt from observing nature. He applied this approach to vast areas of existing knowledge to validate, reject, or add to what was already known in subject areas like physics, philosophy, astronomy and biology. Although a pupil at Plato's Academy for nearly twenty years, the two great thinkers were diametrically opposed on a number of subjects, but Aristotle's theses had just as profound an effect on Western thinking as his master's.

In the area of scientific thought, in particular, Aristotle had an even more fundamental influence, to the point that over the ensuing centuries his proposals were attributed an almost god-like, unchallengeable authority, not always with beneficial results.

EUCLID

C. 330–260BC

CHRONOLOGY Although we possess extensive knowledge about the thoughts of many of the ancients, as we have seen, it is often the case that their lives and times are more obscure; this is certainly true in the case of Euclid. Although a name familiar to every schoolchild, almost nothing is known about his life, when and where he studied, or even when and where he was born and died: a true international man of mystery!

It is said that King Ptolemy I Soter of Egypt asked Euclid if it was possible to master geometry by a more direct route than reading his thirteen volume definitive work on the topic. Euclid famously replied, 'There is no royal road to geometry, your Majesty.' Yet what Euclid had provided was one of the most majestic routes to the subject. It would go on to be revered for over two thousand years.

▸ THE ELEMENTS

Euclid's legacy is well known, and yet, as we

have seen, much of the life of the Greek mathematician remains a mystery. He probably studied under **Plato** at Athens and certainly spent most of his time in Alexandria where he founded a mathematics academy. Whether all the works credited to him, including *Data*, *On Divisions*, the *Optics* and *Phaenomena*, were actually compiled solely by Euclid, or were produced with assistance from students at his school, remains unclear, but the impact of the texts is known to be great. In particular, *The Elements*, Euclid's masterwork on geometry, had a phenomenal influence on

After the Bible, The Elements *has probably been more studied than any other book in history*

Western academic thinking. This is best illustrated by the suggestion that after the *Bible*, *The Elements* has probably been more studied, translated, and reprinted than any other book in history.

The reason for this is twofold: not just what Euclid said, but also the way that he said it. Indeed, the latter is arguably the more enduring of the two because it profoundly influenced the presentation of almost every future mathematical, scientific, theological and philosophical text, amongst others. The reason is because Euclid took a systematic approach to his writing, laying out a set of axioms (truths) at the beginning and constructing each proof of theorem which followed on the basis of the proven truths which had gone before. This logical, 'building block' method set the accepted academic precedent for proving knowledge and continues as a standard model today.

▶ A GEOMETRIC SYNTHESIS

The compilation of knowledge that Euclid brought to the thirteen volumes of *The Elements* was so comprehensive and persuasive that it remained virtually unchanged and unchallenged as a teaching manual for over two millennia. Certainly, many of the theories outlined were not his; he was simply seeking to assimilate all geometric (and much other mathematical) knowledge into a single text. For example, the ideas of previous Greek mathematicians such as Eudoxus, Theaetetus and **Pythagoras** were all evident, though much of the systematic proof of theories, as well as other original contributions, was Euclid's. The first six of the thirteen volumes were concerned with plane geometry, for example laying out the basic principles of triangles, squares, rectangles and circles and any issues around these, as well as outlining other mathematical cornerstones, including Eudoxus's theory of proportion. The next four books looked at number theory, including the celebrated proof that there is an infinite number of prime numbers. The final three works focused on solid geometry.

▶ NON-EUCLIDEAN SPACE

Ironically, it is with some of the text's initial axioms that later mathematicians have found fault. The last axiom in particular has proved to be controversial. This 'parallel' axiom states that if a point lies outside a straight line, then only one straight line can be drawn through the point which never meets the other line in that plane (i.e. the parallel line). This was examined in the nineteenth century by the Romanian mathematician Janos Bolyai. Taking on his father's life work, he attemted to prove Euclid's parallel postulate, only to discover that, in fact, it was unprovable. This began a new school of thought and later, given further weight by Albert **Einstein's** belief that the geometry for space was also non-Euclidean, it was subsequently proved to be true.

THE LEGACY OF EUCLID

Although the discoveries of the last two hundred years have shown time and space to be other than Euclidean under certain circumstances, this should not be seen to undermine his achievements. To have constructed The Elements *in the manner he did, to have had an effect of such magnitude on the development of Western thought, and to have been accepted as the only authority on geometry for so long, (and for most practical purposes still attain such a status) is a profound legacy that few have equalled.*

ARCHIMEDES

C. 287–212 BC

CHRONOLOGY • **213** BC Archimedes' war machines ensure Roman attack on Syracuse is unsuccessful. Siege opens • **212** BC **The** Romans capture Syracuse; Archimedes is killed by a Roman soldier during the sack of the city • **75** BC Archimedes' tomb is discovered and restored by the Roman statesman Cicero

'**G**ive me a place to stand on, and I will move the earth,' Archimedes is reputed to have declared to the people of Syracuse. The practicalities of an earth-bound life may have denied him that particular pedestal but arranging for his patron King Heiron to move a ship by pushing a small lever was considered only a slightly less miraculous feat. With such audacious displays, along with his brilliance as an inventor, mechanical scientist and mathematician, it is no wonder Archimedes was so popular and highly regarded among his contemporaries.

▶ THE MATHEMATICIAN

It was not only his peers, however, who benefited from Archimedes' work. Many of his achievements are still with us today. First and foremost, Archimedes was an outstanding pure mathematician, 'usually considered to be one of the greatest mathematicians of all time,' according to the *Oxford Dictionary of Scientists*. He was, for example, the first to deduce that the volume of a sphere was $4\pi r^3 \times 3$, where r is the radius. Other work in the same area, as outlined in his treatise *On the Sphere and Cylinder*, led him to deduce that a sphere's surface area can be worked out by

'Give me a place to stand, and a long enough lever, and I will move the earth'

multiplying that of its greatest circle by four; or, similarly, a sphere's volume is two-thirds that of its circumscribing cylinder. He calculated pi to be approximately $^{22}/_7$, a figure that was widely used for the next 1500 years.

▸ THE ARCHIMEDES PRINCIPLE

Archimedes also discovered the principle that an object immersed in a liquid is buoyed or thrust upwards by a force equal to the weight of the fluid it displaces. The volume of the displaced liquid is the same as the volume of the immersed object. Legend has it that he discovered this when set a challenge by King Heiron to find out whether one of his crowns was made of pure gold or was a fake. While contemplating the problem Archimedes took a bath and noticed that the more he immersed his body in the water, the more the water overflowed from the tub. He realised that if he immersed the crown in a container of water and measured the water that overflowed he would know the volume of the crown. By obtaining a volume of pure gold equivalent to the volume of water displaced by the crown and then weighing both the crown and the gold, he could answer the King's question. On making this realisation, Archimedes is said to have leapt from his tub and run naked along the street shouting 'Eureka!', 'I have found it!'

▸ LEVERS AND PULLEYS

Indeed, it was the practical consequences of Archimedes' work which mattered more to his contemporaries and for which he became famous.

One such practical demonstration allowed King Heiron to move a ship with a single small lever – which in turn was connected to a series of other levers. Archimedes knew the experiment would work because he had already prepared a general theory of levers. Mathematically, he understood the relationship between the lever length, fulcrum position, the weight to be lifted and the force required to move the weight. This meant he could successfully predict outcomes for any number of levers and objects to be lifted.

Likewise he came to understand and explain the principles behind the compound pulley, windlass, wedge and screw, as well as finding ways to determine the centre of gravity in objects.

▸ ARCHIMEDES GOES TO WAR

Perhaps the most important inventions to his peers, however, were the devices created during the Roman siege of Syracuse in the second Punic War. The Romans eventually seized Syracuse, due to neglect of the defences, and Archimedes was killed by a Roman soldier while hard at work on mathematical diagrams. His last words are reputed to have been, 'Fellow, do not disturb my circles!'

FURTHER ACHIEVEMENTS

Inventions

- Archimedes' Screw: *a device used to pump water out of ships, and also to irrigate fields.*
- Archimedes' Claw: *a huge war machine designed to sink ships by grasping the prow and tipping them over, used in the defense of Syracuse.*
- Compound pulley systems: *enabled the lifting of enormous weights at a minimal expenditure of energy.*
- The method of exhaustion: *an integral-like limiting process used to compute the area and volume of two-dimensional lamina and three-dimensional solids.*

Discoveries

- *Archimedes was responsible for the science of hydrostatics, the study of the displacement of bodies in water (see Archimedes' Principle). He also discovered the principles of static mechanics and pycnometry (the measurement of the volume or density of an object).*
- *Known as the 'father of integral calculus', Archimedes' reckonings were later used by, among others, Kepler, Fermat, Leibniz and Newton.*

HIPPARCHUS

C. 170–125BC

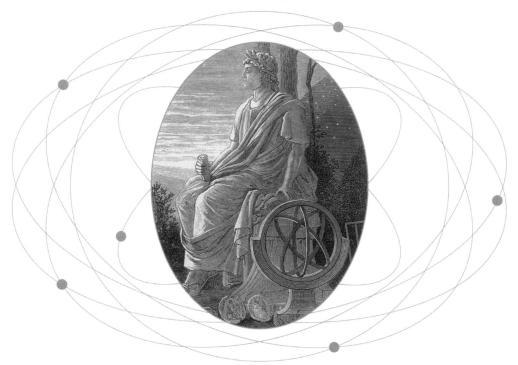

CHRONOLOGY The most significant date pertaining to Hipparchus is 134 BC, when he observed a new star in the constellation of Scorpio. Most of the detail of Hipparchus's life that has come down to us is taken from Ptolemy's record of his achievements. The vast majority of Hipparchus's original work has been lost. He was born in Nicaea, Bithynia, now in modern Turkey, where he undertook some of his astronomical observations, along with sustained periods in Rhodes and to a lesser extent in Alexandria.

Hipparchus spent long periods taking measurements of the earth's position in relation to the stars. The results enabled him to make a number of important findings and calculations.

▶ THE 'PRECESSION OF THE EQUINOXES'

He discovered what is now known as the 'precession of the equinoxes' by comparing his own observations with those noted by Timocharis of Alexandria a century and a half previously together with earlier recordings from Babylonia. What Hipparchus realised was that even taking into account any observational errors made by his predecessors, the points at which the equinox (the two occasions during the year when day and night are of equal length) occurred seemed to move slowly but consistently from east to west against the backdrop of the fixed stars. He gave a value for the annual precession of around 46

The first person to use the concepts of longitude and latitude in his geographical positionings

seconds of the arc, which is exceptionally close to the modern figure of 50.26 seconds, given the tools and data then available to him.

▶ THE DISTANCE OF THE MOON

From these observations, Hipparchus was able to make much more accurate calculations on the length of the year, producing a figure that was accurate to within six and a half minutes. He was also able to correctly determine the lengths of the seasons and offer more exact predictions of when eclipses would take place. He made observations of the sun's supposed orbit and attempted to do likewise with the more irregular orbit of the moon. Although partially successful, he could not make entirely accurate calculations. Using measurements and timings related to the earth's shadow during eclipses, other attempts were made to determine the size of the sun and moon and their distances from the earth. Again, while not entirely accurate, Hipparchus proposed that the distance of the moon from the earth was 240,000 miles. This is remarkably close to the modern figure.

▶ A CATALOGUE OF STARS

Perhaps Hipparchus' most important astronomical achievement was his plotting of the first known catalogue of the stars, despite warnings from some of his contemporaries that he was thus guilty of impiety. He was inspired to begin this work in 134 BC after allegedly seeing a 'new star' which prompted his speculation that the stars were not fixed as had previously been thought. He went on to record the position of 850 stars in the remaining years of his life, a significant achievement given the resources available to him. Moreover, he devised a scale for recording a star's magnitude or brightness: from the most visible – the first magnitude – to the faintest – the sixth. Though amended considerably, it is a scale still used today.

▶ DEVELOPING TRIGONOMETRY

Because of the accelerated developments Hipparchus was making in astronomy, he was required to break new ground in other disciplines, particularly mathematics, to facilitate his celestial observations and calculations. Most notably of all, he developed an early version of trigonometry. With no notion of sine available to him, he was required to construct a table of chords which calculated the relationship between the length of a line joining two points on a circle and the corresponding angle at the centre.

FURTHER INFLUENCE OF HIPPARCHUS

Although Hipparchus is considered to be one of the most influential astronomers of the ancient world, it is arguable that his most impacting achievements lay in the areas of mathematics and geography. The geographer and astronomer Ptolemy cited Hipparchus as his most important predecessor and he is most often revered for his astronomical measurements and cataloguing. Yet as the attributed inventor of trigonometry, as well as being the first person to plot places on the earth's surface using longitude and latitude, his influence was long lasting and widespread.

He was able to apply his work on the trigonometry of spheres to the planet from which he made his observations. Significantly, he was the first person to use longitude and latitude in his mathematical calculations to position where places were on the earth's surface. Like so many of Hipparchus's achievements, it is his further pioneering work that still resonates today.

ZHANG HENG

78–139 AD

CHRONOLOGY • **78** AD Zhang Heng born in Nan-yang, China • **123** AD Corrects the calendar, bringing it into line with the seasons • **132** AD Invents the first seismograph for measuring earthquakes • **138** AD Zhang Heng's machine detects the location of an earthquake 500 kilometers away

Western science is often credited with discoveries and inventions which have been observed in other cultures centuries before. This can be due to a lack of reliable records, difficulty in discerning fact from legend, problems in pinning down a finding to an individual or group, or frequently, simple ignorance. No such excuses exist for the work of Zhang Heng, whose life and achievements are well recorded, and whose major invention was created some 1,700 years before European scientists 'invented' the same thing.

▶ STUDYING THE EARTH

Zhang, a Chinese scholar in the East Han Dynasty, was a man of many disciplines, including astronomy, mathematics and literature. Yet his greatest achievement was in geography, inspired by one of the duties assigned to him in the course of his work as Imperial Historian! China regularly suffered from earthquakes and as part of his job Zhang was required to record when and where they occurred. Rather than accept the common superstition that the quakes were punishment from angry gods, Zhang believed that if he took a scientific approach to

He devised the world's first seismograph, which he named Di Dong Yi, Earth Motion Instrument

noting data about tremors, the Dynasty would be better equipped to predict, prepare for and deal with them. To this end he devised the world's first seismograph, an invention he named Di Dong Yi, or 'Earth Motion Instrument'.

▶ THE EARTH MOTION INSTRUMENT

The seismograph was large, at almost two metres in diameter, and made out of bronze. Eight thin copper rods were attached to a central shaft at one end and to a corresponding number of dragons' heads at the other. These heads pointed in the eight major directions of a compass (north, north-east, east, south-east and so on), and each contained a copper ball in its mouth. Underneath each dragon was an open-mouthed copper frog. When a tremor occurred, the copper ball fell out of the mouth of the dragon nearest to the direction from which the earthquake came and into the frog's mouth, which in turn rang a bell alerting the royal household. A story is recorded that in 138 AD a copper ball had dropped to the west. Zhang recounted his finding to the emperor, but for two days nothing unusual happened and there were no reports of activity elsewhere. Sceptics were left to question the validity of Zhang's machine. Finally, though, messengers arrived on horseback reporting a severe earthquake 500 kilometres to the west. Zhang was vindicated.

▶ OBSERVING THE STARS

Fabled to be a man with intense powers of concentration, Zhang was also able to employ his abilities to excellent effect in astronomy. Through his observations he correctly deduced that the sun caused the illumination of the moon, and that lunar eclipses were caused by the earth's shadow passing over its surface. He mapped the night sky in fine detail, recording 2500 'brightly shining' stars in 124 constellations, 320 of which were named. He estimated that in total, including the 'very small,' there were 11520 stars. In addition, Zhang wrote a number of books on astronomy, the most famous being *Lin Xian*. In another, *Hun-i-chu*, he outlined his perception of the universe and the earth's position within it. 'The sky is like a hen's egg,' he wrote, 'and is as round as a crossbow pellet; the earth is like the yolk of the egg, lying alone at the centre. The sky is large and the earth is small.'

Zhang Heng then, in common with his Greek predecessors, believed that the earth was spherical and at the centre of the universe. This drove him to create possibly the first three-dimensional model of the cosmos: a bronze celestial orb which turned by water-power. Each year, making a single complete rotation, it showed how the stars' positions changed.

FURTHER ACHIEVEMENTS

Zhang undertook other work which had a lasting impact. He improved the previous figure of π from 3, the traditional figure in use by the Chinese, to ÷10 or 3.162, closer to the number of 3.142 used today. Zhang also performed calculations involving time, notably correcting the Chinese calendar in 123 AD to harmonise it with the seasons.

Zhang's seismograph is recognized by the world as an instrument that was well ahead of its time. To this day no one has been able to reproduce it. He constructed the first accurate odometer, or 'mileage cart'.

He is considered one of the four great painters of his era. Zhang also produced over twenty famous literary works.

PTOLEMY

90–168AD

A NOTE ON DATES For all that is known of Ptolemy's work, very little is known about his life. Of Greek descent, he was born and lived in Alexandria, Egypt. It is thought that he rarely, if ever, left Alexandria, ironic for a man who mapped the world. Instead, he obtained his geographical knowledge from the accounts of sailors and Roman visitors.

Claudius Ptolemy's work in astronomy and geography had a profound impact on man's perception of the world and universe from the second century AD until the Renaissance. His genius lay in his ability to distil and summarise the important findings of his predecessors, then add to or provide 'scientific' proof of their theories from an all-encompassing viewpoint. Ptolemy's texts were written with such authority that later generations struggled for more than a thousand years to convincingly challenge his theses.

▸ PTOLEMY'S ALMAGEST

The work for which Ptolemy was most revered is his thirteen-volume *Mathematical Collection*, later more commonly referred to as the *Almagest*. It provided for the first time a definitive compilation of everything that was known and accepted in the field of astronomy up to that point. In particular, the work of **Hipparchus** was used as the starting point of many of Ptolemy's developments and it is largely through the records of the latter that Hipparchus's theories have been passed down to us today. In addition, as the starting point for his

The 'Ptolemaic' astronomical system would not be rivalled until Copernicus, 1,400 years later

arguments, Ptolemy used the **Aristotelian** notion that the earth was at the centre of the universe, with the stars and planets rotating in perfect circles around it. He then set about attempting to justify this interpretation through astronomical observation and mathematical speculation. The result was the 'Ptolemaic System', a mathematically 'proven' interpretation of the universe that would not be rivalled until 1543 by **Copernicus**.

In order to explain his observations in the context of a geocentric model of the universe – which would ultimately be proved wrong – Ptolemy had to introduce some complex explanations and calculations to arrive at a convincing result. Most notably, in explaining planetary and star motion he argued for a system of 'deferents', or large circles, rotating around the earth, and eighty 'epicycles', or small circles, which circulated within the deferents. He also examined theories of 'movable eccentrics'. These proposed just one circle of rotation, with its centre offset slightly from the earth, as well as 'equants' – imaginary points in space which also helped define the focal point of the rotation of the celestial bodies. Ptolemy needed to employ these complex theories because he did not know that the planets actually moved in elliptical orbits, not the perfect circles he supposed. As a result, his predictions for some of their movements continued to be inaccurate but they were the best explanations available at the time, and for many centuries afterwards.

▶ GEOGRAPHY

Almost as significant in its impact on the world was Ptolemy's *Geography*. For the first time, a detailed mathematical explanation for calculating lines of longitude and latitude was offered. and again, this built on the work Hipparchus had begun. It allowed Ptolemy to undertake groundbreaking research into the projection of the earth's sphere onto a plane, leading to the drawing of a scaled map of the known world which would resonate for as long as *Almagest*.

Although there were many errors on this map, such as the equator being too far north and Asia stretching too far to the east, its importance to later generations cannot be underestimated. Most notably, it has been argued that because Asia appeared much closer to Europe on Ptolemy's map than it actually was (assuming the earth was a sphere and one 'looked' westwards), it was this that encouraged Christopher Columbus to sail west in the hope of finding a shorter route to Asia, and to accidentally discover America.

▶ ASTROLOGY

Still frequently read and referred to today, Ptolemy's other major text is his *Tetrabiblos*, possibly the founding work on the then 'science' of astrology. Although this more properly belongs in the category of 'pseudo-science', Ptolemy does at least suppose that the influence of the stars on humans may be due to some sort of radiation.

FURTHER INFLUENCE OF PTOLEMY

Ptolemy wrote on a number of other subjects, and two works in particular were of some importance. His final text is the Optics, *regarded by many as his most successful. In this work, Ptolemy gives a statement of various elementary principles of optics, which he then* sets out to demonstrate. After setting out the *principles of reflection, Ptolemy then proceeds to examine the refraction of rays of light passing through water, providing tables for various angles of incidence, tables which are clearly based upon empirical observation.*

GALEN OF PERGAMUM

130–201AD

CHRONOLOGY
- **129 AD** Galen born in Pergamum (now Bergama in Turkey)
- **148–157AD** Travels and studies in Corinth and Alexandria
- **157AD** Takes the post of surgeon to the Pergamum gladiators • **c. 161 AD** Becomes physician to the emperors Marcus Aurelius and Commodus • **1628** William Harvey's system of blood circulation becomes the first viable alternative to Galen's

Unlike many of the other entrants in this book, Galen is not famous for any single achievement, but more for the sheer volume of medical thought which he presented. That accomplishment in itself may not necessitate an inclusion, but the fact that his works on medical science became accepted as the only authority on the subject for the following 1400 years does.

▶ **UNDISPUTED FOR A MILLENNIA**

The question, therefore, is why? Some commen-taries suggest the answer is simply because Galen's studies were so all-encompassing that there was very little left for those following him to dispute. Another is the readiness with which the Arab, Christian and Jewish authorities accepted his work, lending it a weight which might have made it difficult for others to challenge. A third explanation could be that Galen not only incorporated the results of his own findings in his texts, but also compiled the best of all other medical knowledge that had gone before him into a single collection, such as that of

Galen's studies were so comprehensive, those following him had very little to dispute

Hippocrates, for example. In particular, Galen readily adopted Hippocrates' 'four humours' approach to the body, and this was one of the main reasons it endured for so long.

▶ A METICULOUS INVESTIGATOR

That is not to say Galen was at all lacking in original material and thinking. He was meticulous and methodical in his approach to his own medical investigations, above all in anatomy. Many important dignitaries came to the shrine of Asklepios, the god of healing in Galen's home town, to seek cures for ailments. Thus Galen was able to observe first hand the symptoms and treatments of diseases. After spells in Smyrna (now Izmir), Corinth, and Alexandria studying both philosophy and medicine, which he considered inextricably linked, and including work on the dissection of animals, he returned to Pergamum in 157. There he took up a four-year appointment as a physician to gladiators, giving him further first hand experience in practical anatomical medicine.

▶ PHYSICIAN TO EMPERORS

All of this was excellent preparation for his transfer to Rome. Here he spent most of the rest of his career and became the esteemed physician to emperors Marcus Aurelius, Lucius Verus, Commodus and Septimius Severus. This position not only brought him prestige, but it allowed him the freedom to undertake detailed research and dissection in the quest for the improved knowledge it provided. Galen was not permitted to scrutinise human cadavers, so he dissected animals, predominantly Barbary apes because of the characteristics they shared with man. His most influential conclusions concerned the central operation of the human body. Sadly they were only influential in that they limited the search for accurate information for the next millennia and a half.

Galen believed that blood was formulated in the liver, the source of natural spirit. In turn, this organ was nourished by the contents of the stomach which was transported to it. Veins from the liver carried blood to the extremes of the body where it was turned into flesh and 'used up', thus requiring more food on a daily basis to be converted into blood. Some of this blood passed through the heart's right ventricle, then seeped through to the left ventricle and mixed with air from the lungs, providing vital spirit which regulated the body's heat and blood flow. Using the arteries, a portion of this blood was then transported to the brain where it blended with animal spirit. This created movement and the senses. The combination of these three spirits managed the body and contributed to the make up of the soul. It was for this reason that Galen missed the idea of a single, integrated system of the circulation of the blood, a result which was not conclusively proved until 1628 by William Harvey.

GALEN'S LEGACY

Although some of Galen's deductions were wrong, his surviving 129 volumes are a phenomenal contribution to his subject and offered a platform from which Renaissance physicians could begin their critical progress. It was Galen who first introduced the notion of experimentation to medicine.

Many of the anatomical errors made by Galen were due due to the fact that he could only operate on animals – human dissections were out of favour at the time. Galen became a doctor supposedly because his father had a dream in which Asklepios, the god of healing, appeared to him.

AL-KHWARIZMI

800–850

CHRONOLOGY • c. 786 Al-Khwarizmi born in Khwarizm, now Khiva, in Uzbekistan • 813 Caliph al-Ma'mun, the patron of Al-Khwarizmi, begins his reign in Baghdad • c. 820 'House of Wisdom' founded in Baghdad by al-Ma'mun • 833 Death of al-Ma'mun

One of the greatest scientific developments of all time was the introduction of 'Arabic' numerals into mathematics. The man often credited with their invention was Al-Khwarizmi, an Arabian mathematician, geographer and astronomer. Yet strictly speaking, the concept was neither invented by Al-Khwarizmi, nor was it Middle Eastern in origin.

▸ NUMERICAL NOTATION

What Al-Khwarizmi did do for Arabic numerals, though, was introduce them to Europe, which is why many Western textbooks subsequently acknowledged the development as his. The notation actually finds its roots in India around 500 AD and the naming of the numerical scheme employing the figures, now called the 'Hindu-Arabic' system, acknowledges this. The method of using only the digits 0–9, with the value assigned to them determined by their position (e.g. the '1' in '100' has a different value than the '1' in '10' because of its location in relation to other digits), as well as introducing a symbol for zero for the first time, completely revolutionised mathematics. Without it many of the develop-

The name 'Al-Khwarizmi' became 'algorithm', meaning 'rule of calculation', in the West

ments of later times, and what have become norms in the modern world, would have been impossible. Al-Khwarizmi observed this system, then clearly explained how it worked in his text *Calculation with the Hindu Numerals*. When translated later into Latin, it was widely adopted by the West and ultimately the entire world. Even today the numerical system is perhaps the only truly global 'language'.

▸ THE HOUSE OF WISDOM

Al-Khwarizmi does have a much more original claim, however, to writing the first book on algebra, and indeed, introducing the word into our language. He was afforded the opportunity to develop such texts as a patron to Caliph Al-Ma'mun in Baghdad, who ruled the huge Muslim empire extending from the Indian subcontinent to the Mediterranean. Al-Ma'mun's father, Caliph Harun al-Rashid, had been keen to facilitate the development of academic disciplines in his kingdom, and Al-Ma'mun had continued to back his father's goals, founding his 'House of Wisdom' to this end. This academy housed a library, including translations of important Greek texts, and also established astronomical observatories. Al-Khwarizmi repaid the investment with his work *Calculating by Completion and Balancing*.

▸ A PRACTICAL GUIDE TO ARITHMETIC

In this treatise, Al-Khwarizmi set out to provide a practical guide to arithmetic using, where applicable, calculations later described as algebraic. In so doing he introduced quadratic equations, although he described them in words and did not use the symbolic algebra (e.g. $x^2+3x=10$) we more commonly understand today. The two key concepts he outlined were the ideas of 'completion' and 'balancing' of equations. Completion is the method of expelling negatives from an equation by moving them to the opposite side (e.g. $4x^2=54x-2x^2$ becomes $6x^2=54x$). Balancing, meanwhile, is the reduction of common positive terms on both sides of the equation to their simplest forms (e.g. $x^2+3x+22=7x+12$ becomes $x^2+10=4x$).

▸ THE 'FATHER OF ALGEBRA'

It is not clear whether Al-Khwarizmi was familiar with the works of Euclid, despite the fact that one of his colleagues at the House of Wisdom had translated *The Elements* into Arabic.

Although Al-Khwarizmi was clearly building on the work of others before him, such as Diophantus and Brahmagupta, his was a much closer expression of modern elementary algebra, and this is the reason he is sometimes referred to as 'the father of algebra.'

Indeed, the Arabic title of *Calculating by Completion and Balancing* is *Hisab al-jabr w'al-muqabala* and it is from the 'al-jabr' in this heading that the word 'algebra' descends.

FURTHER ACHIEVEMENTS

Through carelessness of pronunciation the name 'Al-Khwarizmi' became referred to in the West as 'algorismi', then 'algorism', and ultimately 'algorithm', which is where we get the word meaning 'a rule of calculation' today.

Al-Khwarizmi undertook other work in mathematics, such as writing tables of sines

and tangents. He also performed many astronomical observations and was a keen investigator of geography. In particular, he expanded on Ptolemy's use of longitude and latitude in plotting the positions of places around the world, developing a series of maps more accurate than those of his predecessor.

JOHANNES GUTENBERG

1400–1468

• 1420 Gutenberg moves from Mainz, Germany, to Strasbourg, France • 1450 Returns to Mainz, and sets up his printing press using moveable type • 1450–56 Prints a number of books, a calendar, and a letter of Papal indulgence • 1456 prints his famous 42-line Bible • 1465 Made courtier to the Archbishop of Mainz

Johannes Gutenberg was born and spent much of his life in the German town of Mainz. His family background was in minting and metal-working, an ideal foundation for his training as an engraver and goldsmith. These skills enabled him to craft the first individual metal letter moulds, matrices, which were the core of his achievements in printing. Hand-held block printing, however – a laborious process of carving whole pages of 'fixed' text out of wooden slabs, reproducing copies using dyes – had been used for many decades before the German inventor appeared.

▶ MOVEABLE TYPE

What Gutenberg mastered was the idea of placing individual metal letters into temporary mounts, which could then be dismantled or 'moved' once a page of text had been completed and reused to produce other pages. In comparison to the slow engraving and single use of wooden blocks, the theoretically infinite number of different sides which could be made out of a set of metal characters, together with the speed at which a template could be created, revolutionised printing and the spread of the printed word.

The development closest to the impact of the numerical system in the history of science

It is thought that Gutenberg began experimenting with the creation of metal letter casts towards the end of the 1430s during a period living in Strasbourg. It was probably not until sometime between 1444 and 1448, however, that he finally perfected the development of the moveable type printing press. Certainly, it is known he borrowed money from a relative in 1448 on his return to Mainz most probably to fund his printing business. The invention itself consisted of an adapted wine press with the plate of metal characters at the bottom upon which a piece of paper was laid and the top of the press lowered from above to force the imprint. The matter of developing a suitable dye for this machine was no easy task, either, but it is thought Gutenberg finally found the answer in an amalgamation of linseed oil and soot.

▶ THE FORTY-TWO LINE BIBLE

No works survive bearing Gutenberg's name, but the earliest printed piece attributed to him is a Calendar for 1448. Much more famous, with around forty-eight of the original two hundred copies still in existence, is the first Bible printed using moveable type, known as the forty-two line Bible' because of the number of lines to a page. It is believed Gutenberg and his assistants made the copies between 1450 and 1456.

In the later years of his life, Gutenberg lived off the patronage of the Archbishop of Mainz, an offer thought to be an acknowledgement of his ground-breaking achievement. Others, however, have been less willing to recognise Gutenberg as the inventor of moveable type and instead claim the inventor of printing to be a certain Laurens Janszoon Coster (c. 1370–1440).

Very little is known about this Dutchman and, like Gutenberg, no printed fragments bearing his name survive, but a legend persists that he carved individual letters into wood for the entertainment of his grandchildren. To amuse them further he used dye to print words and sentences onto paper whereupon he realised the possibilities for these moveable pieces. A block printer by trade, it is thought that Coster began using the wooden characters to speed up his printing processes probably by bringing together a combination of block and moveable type in the production of texts. The evidence to support these claims is limited at best. Even if true, what is notable is the superior quality of Gutenberg's metal casts and press – they are almost as important as the idea of moveable type itself.

Some sources credit the Chinese with inventing moveable type printing, using characters made of wood, in the early fourteenth century. It is almost certain, however, that Gutenberg developed his ideas independently and was unaware of any similar developments which may or may not have taken place on the other side of the world more than a century before him.

THE LEGACY OF GUTENBERG

The one development in the history of science which probably comes closest to matching the impact of the Hindu-Indian numerical system is the invention of the printing press using moveable type. Although not specifically a scientific achievement, its emergence provided one of the key tools in helping to begin the revolutionary progress of the subject in Europe, giving academics the opportunity to share scientific knowledge widely and cheaply.

By the end of the fifteenth century tens of thousands of books and pamphlets were already in existence and the stage was set for the imminent explosion of scientific ideas.

LEONARDO DA VINCI

1452–1519

CHRONOLOGY • **1469** Leonardo apprenticed to the studio of Verrocchio in Florence • **1482** Works for the Duke of Milan • **1502** Returns to Florence to work for Cesare Borgia as his military engineer and architect • **1516** Journeys to France on invitation of Francis I • **1519** Dies in Clos-Luce, near Amboise, France

It is perhaps something of an indulgence to include Leonardo Da Vinci in a book of scientists who changed the world, not least because most of his work remained unpublished and largely forgotten centuries after his death. His, however, was undoutedly one of the most brilliant scientific minds of all time; arguably the biggest handicap preventing him from profoundly changing the world was the era in which he lived.

The genius of Leonardo's designs for his inventions so far outstripped both his contemporaries' intellectual grasp and contemporary technology that they were rendered literally inconceivable to anyone but him. If Leonardo could have teleported to **Edison's** time, with his access to nineteenth century technology, one can only speculate how much more he may or may not have achieved than even Edison himself. But even in his own time, Leonardo's achievements were notable.

▶ RENAISSANCE MAN

Leonardo is celebrated as the Renaissance artist who created such masterpieces as the Last Supper (1495–97) and the Mona Lisa (1503–06), yet

The term 'Renaissance Man' could have been coined specifically for Leonardo

much of his time was spent in scientific enquiry, often to the detriment of his art. The range of areas Leonardo examined was breathtaking. It included astronomy, geography, palaeontology, geology, botany, zoology, hydrodynamics, optics, aerodynamics and anatomy. In the latter field in particular, he undertook a number of human dissections, largely on stolen corpses, to make detailed sketches of the body. Irrespective of the breadth of his studies, however, perhaps the most important contribution Leonardo made to science was the method of his enquiry, introducing a rational, systematic approach to the study of nature after a thousand years of superstition. He would begin by setting himself straightforward scientific queries such as 'How does a bird fly?' Next he would observe his subject in its natural environment, make notes on its behaviour, then repeat the observation over and over to ensure accuracy, before making sketches and ultimately drawing conclusions.

▶ AERODYNAMICS

Moreover, in many instances he could then directly apply the results of his enquiries into nature to designs for inventions for human use. For example, his work in aerodynamics led him to make sketches for a number of flying machines, – which, potentially, could have flown – including a primitive helicopter, some five hundred years before the invention became reality! He even envisaged the need for his flying machines to have a retractable landing gear to improve their aerodynamics once airborne. In 1485 he designed a parachute, three hundred years before it became an actuality, and included calculations for the necessary size of material to safely bring to ground an object with the same weight as a human. He also had an excellent understanding of the workings of levers and gears, enabling him to design bicycles and cranes.

▶ HYDRODYNAMICS

Leonardo's studies in hydrodynamics led to numerous sketches on designs for waterwheels and water-powered machines centuries before the industrial revolution. In addition, he sketched humidity-measuring equipment as well as a number of primitive diving suits, mostly with long snorkel devices to provide a supply of air.

▶ MILITARY INVENTIONS

During his work for the Duke of Milan between 1482 and 1499, Leonardo prepared an array of designs for weaponry such as catapults and missiles. Even in this arena, however, he could not help but create sketches of weapons that lay way ahead of their time such as hand-grenades, mortars, machine-type guns, a primitive tank and, most audaciously, a submarine!

LEONARDO'S INFLUENCE ·

If this were a book of scientists who 'could have' changed the world, Leonardo Da Vinci would be at the top of the list. But regardless of the fact that many of the designs for his potentially world-changing creations were never published, his methodical approach to science marks a significant and symbolic stepping-stone from the Dark Ages into the modern era.

Hoping to secure employment with the Duke of Milan, he wrote to him that his areas of expertise included: the construction of bridges and irrigation canals, the designing of military weapons and architecture, as well as painting and sculpture. To add to the list, Leonardo is also credited with being the first ever person to conceive of a bicycle!

NICOLAS COPERNICUS

1473–1543

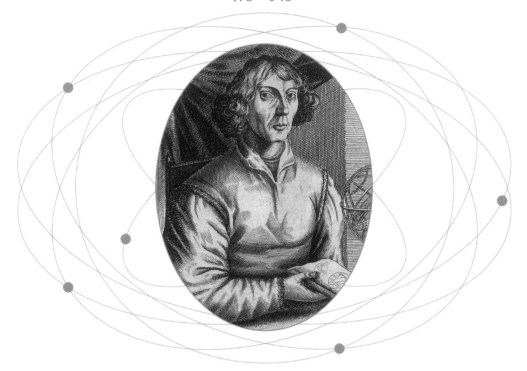

CHRONOLOGY • 1491 Copernicus enters the University of Krakow • 1510–14 The revolutionary *Commentariolus* is circulated • 1543 *De revolutionibus orbium coelestium* (*On the Revolution of the Celestial Spheres*), his definitive work, is published while he is on his deathbed, but is banned by the Catholic Church. The ban is not lifted until 1835.

For all the impact the idea the planets might revolve around the sun, not the earth, would have on astronomy and science, arguably its biggest challenge would be to religion. The explanation of an earth inhabited by human beings, made in God's image as the most superior of all creatures, at the centre of a cosmos around which everything else revolved, suited the Christian Church's inter-pretation of the universe and mankind's position within it. It was a concept which dated back to **Aristotle**, was given observational legitimacy by **Ptolemy** and authority by Christendom. The

Catholic religion still opposed the heliocentric model of planetary motion nearly three centuries after it was first published. And yet ironically its author, Nicolaus Copernicus, was himself a man of the Church.

▸ A MAN OF FAITH

Indeed, it was Copernicus's faith which had led him to question Ptolemy's accepted geocentric model of the universe in the first place. Why would God create a hugely complicated system of equants, epicycles and eccentrics, as Ptolemy had proposed, to explain the planets' motion around

Copernicus literally used the Church to advance his studies, observing the stars from a belltower

the earth when it would be far more simple, logical and graceful to have them all revolving around the sun? It was a theory Copernicus spent many years contemplating while studying in Krakow and then Italy, and continued to develop as he returned to Poland to take up a post as canon in Frauenberg Cathedral. He even used his position within the Church to quite literally advance his studies, using a cathedral tower to quietly and solitarily observe the stars.

▸ THE EARTH CIRCLES THE SUN

Gradually Copernicus became more convinced of his proposition that a fixed sun was at the centre of planetary motion, with the earth rotating around it once a year. Between 1510 and 1514 he drafted *Commentariolus*, his initial exposition of the theory. In order to have any credence, the idea also required that the earth itself was not fixed in position as had previously been thought, but revolved on its axis once every twenty-four hours. This would also explain the apparent movement of the stars and sun across the sky. Perhaps because of his position within the Church, fearing a backlash, or perhaps because he was a perfectionist and recognised that his ideas were not fully developed, Copernicus resisted publishing *Commentariolus*, circulating it instead only among friends.

▸ CHURCH OPPOSITION

Copernicus continued to work on his ideas for the next twenty years and though his final work was largely completed by 1530 he continued to resist pleas by his friends to publish. Word of Copernicus' theories was already spreading across Europe and it is thought that even the Pope himself knew of them but offered no initial resistance to the idea of a heliocentric model. Indeed, it was not until 1616 that the Church banned the text Copernicus eventually published for its 'blasphemous' content, although that sanction subsequently remained in place until 1835, long after the 'Copernican system' had been widely accepted by most others.

▸ A CRITICAL RECEPTION

On The Revolutions of Celestial Spheres was finally published in 1543. But as powerful and revolutionary as Copernicus's theories were, the text was rejected by many academics. This was partially because the author had undermined the simplicity of his initial ideas by clinging onto the Aristotelian belief that planetary motion took place in perfect circles. As we now know this not to be true, it meant Copernicus had been forced to introduce his own system of epicycles and other complex motions to fit in with observational evidence, thereby producing as equally complicated an explanation as the geocentric one he had initially rejected for its lack of simplicity. It was not until Johannes **Kepler** offered the solution that the planets moved in an elliptical, not circular, motion in 1609 that the simplicity Copernicus had been seeking was offered and the rest of his model could be vindicated.

A MAN OF CONTRADICTIONS

Copernicus was brought up by his maternal uncle Lucas, the Bishop of Ermeland, and took a doctorate in canon law at the University of Ferrara in 1503. By this time he had become a canon of Frauenburg. Throughout his life

Copernicus struggled to come to terms with the conflict between his mathematics and his religious faith. Indeed, one of the main reasons he did not publish his works was through fear of contradicting the Bible.

ANDREAS VESALIUS

1514–1564

CHRONOLOGY • 1514 Born in Brussels, Belgium • 1537 Appointed Professor of Anatomy and Surgery, Padua University • 1543 Publishes first anatomically accurate medical textbook, *De humani corporis fabrica* (*On the Structure of the Human Body*) • 1543 Joins Hapsburg court where he serves as physician to the Emperor Charles V and King Philip II of Spain • 1564 Dies on a pilgrimage to the Holy Land

It takes a brave person to challenge the accepted authority on any subject, especially one which has endured without dispute for some 1400 years, and more especially when the person raising the objection is only twenty-eight years old and has only relatively recently graduated. That is just the task Andreas Vesalius took upon himself, however. For many of his contemporaries, there was nothing about this confrontation to consider as 'brave': instead, they described him as anything from a liar to a madman.

▶ CHALLENGING GALEN

The authority Vesalius dared to challenge was that of **Galen**, the celebrated Roman physician who wrote what had been considered the definitive work on human anatomy. Such was his clout that when dissections of humans began to be permitted in Europe from the fourteenth century for research and tuition purposes, lecturers would simply read directly from Galen as the cadaver was cut by a butcher or assistant. Yet what was somehow lost sight of in all of this reverence was the fact that Galen himself had never actually

Vesalius encouraged the hands-on approach, revolutionising the teaching of anatomy

dissected a human body, forbidden as this had been by Roman religious laws. Academics before Vesalius, however, still considered Galen as the authority on the subject, with any advance on his texts regarded as impossible.

▶ A NEW APPROACH

Vesalius's approach was completely different. Born and raised in Belgium, to a family with a distinguished background as doctors to royalty, Vesalius was a keen dissector of animals from a young age. He went on to study medicine at institutions around Europe, notably the universities of Louvain, Paris and then Padua, where he was appointed Professor of Anatomy and Surgery at the age of 24. He insisted on performing the dissection of human bodies himself during lectures to students, rejecting the traditional clean-handed, textbook method.

▶ HUMAN ANATOMY

Although schooled in the Galenic tradition like all other medical students, Vesalius began questioning its teachings towards the end of the 1530s. From 1540 onwards, having been granted an ample number of human corpses to dissect, mostly from the local executioners, Vesalius was convinced. Galen's findings, he argued, did not reflect the human anatomy, but that of apes. This had led to numerous errors based on assumptions Galen had made on similarities between the two.

▶ THE DEFINITIVE TEXT

In 1543, Vesalius published his masterwork *De Humani Corporis Fabrica Libri Septem* or *The Seven Books on the Structure of the Human Body*. It was the first definitive work on human anatomy actually based on the results of methodical dissections of humans and, as such, was the most accurate work on the subject ever written. Furthermore, it was beautifully and clearly illustrated with woodcut drawings, probably prepared at the artist Titian's studios, and was excellently structured and organised. Its publication outdated all that had gone before and the text became the guide upon which future teachers would base their lectures. It was some time before its wisdom was widely accepted, however, due to the hostility which Vesalius often endured for challenging Galen. For example, Vesalius stated he could find no evidence for the 'pores' in the heart which allowed blood to seep from the right to the left ventricle, a key foundation of the Galenic tradition and one which was resolutely defended by many of his contemporaries.

Vesalius spent much of the rest of his life after the *Fabrica* in the service of kings, firstly as the physician to Charles V, the Holy Roman Emperor, then to Phillip II of Spain. He left Spain in 1564 on a pilgrimage to Jerusalem, but died on the return journey.

FURTHER ACHIEVEMENTS

In spite of his premature death, Vesalius left behind a revolutionary legacy to anatomy students. It was only after his publications that both anatomy and medicine in general were first treated as sciences in their own right. By his reasoned critical approach to Galen, he had broken the reverence ascribed to the former

'master' and created a model for independent, rational investigation for his successors in the development of medical science.

Vesalius also changed the organisation of the medical school classroom, and actively encouraged the participation of medical students in dissection lectures.

WILLIAM GILBERT

1540–1603

CHRONOLOGY • 1569 Receives degree at Cambridge university • 1600 Publishes *De magnete, magnetisque corporibus, et de magno magnete tellure* (*On the Magnet, Magnetic Bodies, and the Great Magnet Earth*), the first great English scientific work. • 1600–03 Serves as physician to Queen Elizabeth

William Gilbert has often been considered one of the first great English scientists and arguably the first great physicist of the modern era. His principle area of study related to magnetism in which he made groundbreaking revelations. For all the fame the subject of his observations brought him, however, his method of enquiry is equally, if not more, significant.

▶ DANGEROUS TIMES

Living in the time of Shakespeare and Elizabeth I,

for whom he acted as physician from 1600–03, England was still largely a place of superstition and religious fervour. Rational scientific enquiry was rare with some of the few earlier European attempts at it, such as the observations of Leonardo **Da Vinci**, unknown to Gilbert. He was, however, familiar with the work of **Copernicus** with whom he passionately concurred, a potentially dangerous sentiment in an era when elsewhere in Europe others such as Giordano Bruno and later **Galileo** were being persecuted (and in the case of Bruno, executed) for sharing the same opinion.

Gilbert's first work, De Magnete, is considered to be one of the first scientific texts

▸ NEW METHODS

Given this background, then, Gilbert's approach to his work is all the more remarkable. In an unheard of manner, he cast aside all prior speculation on his subject, including that of the 'authorities' from antiquity, and resolved to only make deductions based on proof. Although that approach seems perfectly natural to the modern reader, it was a rational mode of enquiry which religion and superstition had hitherto largely made impossible. Gilbert's work was instrumental as a model for the scientific revolution.

By the same token, his principle work, *De magnete, magnetisque corporibus, et de magno magnete tellure* or *On the Magnetic, Magnetic Bodies, and the Great Magnet Earth* (1600), is considered to be one of the first truly scientific texts. It was the result of years of painstaking observations and experiments which Gilbert had undertaken to learn more about magnetism and electricity, a term he popularised, and to systematically dispel common myths. For example, it had been believed garlic could destroy the accuracy of a compass needle, one of many folk tales Gilbert sought to redress.

▸ FROM EXPERIMENT TO CONCLUSION

What he did prove through his repeated experiments was that a spherical magnet would force a small compass needle to point north or south pole-ward according to where it was positioned near the sphere, and also 'dip' downward towards its surface. This mimicked the behaviour of a normal compass needle when used under ordinary circumstances in the wider world. From these results, he deduced that the earth itself was effectively a large magnet, with a magnetic 'bar' running through the centre of it (causing the compass to 'dip'), which contained north and south poles at its extremities. Although, these revolutionary findings were not confirmed beyond doubt for several hundred years, they were a vital discovery in beginning to truly comprehend the physics of the earth and even the wider universe beyond.

▸ INVISIBLE FORCES

Indeed, Gilbert went on to reason that magnetism played a part in holding the planets in their orbits. This established the concept of invisible forces and explained much of the behaviour of the universe, which Galileo and **Newton** would go on to exploit. He also correctly surmised that the earth's atmosphere was not very deep at all and the vast majority of space between planets was a vacuum. Through further observations involving experiments with amber, which was known to cause static electricity, he suggested that there might be some kind of link between electricity and magnetism, a theory which equally was not conclusively proven for several centuries.

FURTHER ACHIEVEMENTS

As well as his insistence upon modern methodology in scientific practice, Gilbert introduced a number of terms into the English language including: magnetic pole, electric force and electric attraction. A term of magnetomotive force, the gilbert, is named after him, and it was he who first popularised the term electricity.

Gilbert also disproved many commonly held beliefs about magnetism, including the belief that a diamond can magnetise iron.

As a further contribution to the study of magnets and magnetism, he proved that the earth acts as a bar magnet with magnetic poles.

FRANCIS BACON

1561–1626

• 1561 Bacon born in London • 1594 Bacon receives MA from Cambridge • 1605 Accession of James I of Scotland to the throne of England • 1607 James I appoints Bacon King's Counsel • 1613 Appointed Attorney General • 1620 His *Novum Organum* insists that the correct method for science is experimentation • 1621 Bacon's legal career ends in scandal and ignominy

I f William **Gilbert** gently suggested a new rational approach to experimental science in his famous book *De magnete*, then Francis Bacon stood on the rooftops and bellowed out its arrival to the world. Although not strictly a scientist himself, Bacon was responsible for crystallising the methodology behind the scientific revolution which would go on to change the world so drastically. Ironically, he knew little about Gilbert's book, but nonetheless implored academics to introduce a systematic approach to their studies, following Gilbert.

▶ AN EARLY START

Bacon proved his academic capability from a very young age, entering Trinity College, Cambridge, at only twelve years old. At the age of twenty-three he became Member of Parliament for Dorset by which stage he had also qualified as a barrister. He would go on to have a prestigious career in the Royal Court of James I, rising to Lord Chancellor of England. Having risen to such heights, however, his fall from grace in 1621 was all the more great when he was convicted of taking a bribe while a judge,

Bacon stood on the rooftops and bellowed out the arrival of the rational method

and was stripped of his office and power. Both during and after his legal career, Bacon undertook academic studies in philosophy and science. In the history of science, he is notable for two texts in particular. *The Advancement of Learning* in 1605 signalled his dissatisfaction with the limits of and approaches to knowledge to date, and foresaw a future where the work of the ancient masters would be far surpassed. The *Novum Organum*, of 1620, advanced this sentiment, boldly challenging the **Aristotelian** view and approach to the world. Aristotle himself had written a text called *Organum* or *Logical Works*, and Bacon's 'new' approach to the work of his predecessor suggested in the title alone an alternative direction to scientific study.

▶ A CRITICISM OF METHOD

In the text itself, Bacon strongly criticised Aristotle's 'deductive' method of science, involving formulating abstract ideas and 'logically' building upon them step-by-step to find 'truths', without thorough consideration of whether the theoretical foundation in itself was ever valid. Alternatively then, Bacon argued for 'inductive' reason where the only 'certain' statements that should ever be made were based on repeated observation and proof collected from the natural world. Rather than rely on superstition or accept unquestioningly the flawed solutions of the ancient academics as had largely been the case for two thousand years, Bacon implored scientists to only draw conclusions from exactly what was 'known'. Gathering the data from which to induce these certainties involved a rational, systematic, scientific approach using Bacon's 'Tables of Comparative Instances', which basically provided a methodology for eliminating irrelevancies when examining any given question, and pinpointing the proven facts.

▶ AN EARLY FINISH

Novum Organum was only one part of what Bacon had envisaged as a six-part work outlining his new approach, to be called *Instauratio*. This was never finished, owing to the premature death of the scholar from bronchitis, but his plan indicated that as well as outlining the new method of scientific enquiry, he had hoped to reclassify the sciences into new divisions, assemble a collection of scientific 'facts', provide proven examples using his new method and prepare and espouse a new philosophy based on the practical success of his approach.

FURTHER ACHIEVEMENTS

Even without the planned additional material that Bacon never got around to writing, his work went on to profoundly influence the science of the future. In particular, the science of the seventeenth century, as the scientific revolution found a framework within which to operate. In many instances it continues to do so today.

Bacon cautioned those trying to practise his new method, urging them to repudiate four kinds of intellectual idol:

- *Perceptual illusions –' idols of the tribe'*
- *Personal biases – 'idols of the cave'*
- *Linguistic confusions – 'idols of the marketplace'*
- *Dogmatic philosophical systems – 'idols of the theatre'*

Only once we have abandoned metaphysical baggage, said Bacon, can we approach the scientific method in the correct manner.

GALILEO GALILEI

1564–1642

CHRONOLOGY • 1564 Galileo Born in Pisa, Italy • 1581 Studies medicine at Pisa, but fails to complete the course • 1583 Observes swinging lamps in Pisa Cathedral and notes that the time for the swing is the same no matter what the amplitude • 1586 Invents a hydrostatic balance for the determination of relative densities • 1610 Designs and constructs a refracting telescope. Publishes observations in *Sidereus nuncius* (*Starry Messenger*) • 1632 *Dialogue Concerning the Two Chief World Systems* published. This leads to Galileo being forced by the Church to recant his Copernican views. He is put under house arrest

In both his life and through the imprisonment which he was forced to endure in the years leading up to his death, Galileo more than any other figure personified the optimism and struggle of the scientific revolution. He was responsible for a series of discoveries which would change our understanding of the world, while struggling against a society dominated by religious dogma, bent on suppressing his radical ideas.

▶ A MATHEMATICIAN

Although he was initially encouraged to study medicine, Galileo's passion was mathematics, and it was his belief in this subject which underpinned all of his work. One of his most significant contributions was not least his application of mathematics to the science of mechanics, forging the modern approach to experimental and mathematical physics. He would take a problem, break it down into a series

'Nevertheless, it turns!'

GALILEO, AFTER BEING FORCED TO RENOUNCE HIS HELIOCENTRIC VIEW OF THE EARTH

of simple parts, experiment on those parts, and then analyse the results until he could describe them in a series of mathematical expressions.

One of the areas in which Galileo had most success with this method was in explaining the rules of motion. In particular, the Italian rejected many of the **Aristotelian** explanations of physics which had largely endured to his day. One example was Aristotle's view that heavy objects fall towards earth faster than light ones. Through repeated experiments rolling different weighted balls down a slope (and, legend has it, dropping them from the top of the leaning tower of Pisa!), he found that they actually fell at the same rate. This led to his uniform theory of acceleration for falling bodies, which contended that in a vacuum all objects would accelerate at exactly the same rate towards earth, later proved to be true. Galileo also contradicted Aristotle in another area of motion by contending that a thrown stone had two forces acting upon it at the same time; one which we now know as 'momentum' pushing it horizontally, and another pushing downwards upon it, which we now know as 'gravity'. Galileo's work in these areas would prove vital to Isaac **Newton's** later discoveries.

▶ THE PENDULUM

Galileo's earliest work involved the study of the pendulum, inspired by observing a lamp swinging in Pisa cathedral. Following further experiments, he concluded that a pendulum would take the same time to swing back and forth regardless of the amplitude of the swing. This would prove vital in the development of the pendulum clock, which Galileo designed and was constructed after his death by his son.

▶ THROUGH THE TELESCOPE

One of the inventions Galileo is often mistakenly credited with today is the invention of the telescope. This is not true; there had been a numerous early prototypes mostly developed in Holland before him, and a Dutch optician called Hans Lippershey applied for a patent on his version in 1608. Galileo did, however, develop his own far superior astronomical telescope from just a description of Lippershey's invention, and quickly employed it to make numerous discoveries. A strong supporter of the **Copernican** view of planetary motion, Galileo's initial findings published in the *Sidereal Messenger* (1610) provided the first real physical evidence to back up this interpretation. As well as discovering craters and mountains in the moon, sunspots and the moonlike phases of Venus for the first time, he also noted faint, distant stars which supported the Copernican view of a much larger universe than **Ptolemy** had ever considered. More importantly, he discovered Jupiter had four moons which rotated around it, directly contradicting the still commonly held view, including that of the Church, that all celestial bodies orbited earth, 'the centre of the universe.'

GALILEO AND COPERNICUS

Galileo's Dialogue Concerning the Two Chief World Systems – Ptolemaic and Copernican *in which the Ptolemaic view was ridiculed, attracted the attention of the Catholic Inquisition when it was published in 1632. Threatened with torture, Galileo renounced the* Copernican System. *His work was placed on the banned 'Index' by the Church where it remained until 1835, and he was subject to house arrest for life. But the tide of scientific revolution Galileo had helped instigate proved too powerful to hold back.*

JOHANNES KEPLER
1571–1630

CHRONOLOGY • 1600 Kepler works in Prague with Tycho Brahe, the imperial mathematician • 1601 On Brahe's death, Kepler inherits his position • 1609 Publishes *Astronomia Nova*, containing two of his laws of planetary motion • 1611 Publishes *Dioptrics,* which has been called the first work of geometrical optics • 1619 Publishes *Harmonices mundi* (*Harmonics of the World*). It contains his third law of planetary motion

The German mathematician, Johannes Kepler, while probably not as well remembered as **Copernicus**, was one of the key reasons why the Polish astronomer's theories finally became widely accepted. What Copernicus had started in suggesting a heliocentric model of the solar system, i.e. that the planets actually rotated around the sun, Kepler finished in providing the arithmetical and observational proof to support such a thesis.

▶ TYCHO BRAHE

Kepler himself ultimately owed much of his success to the most famous astronomer of the second half of the sixteenth century, Tycho Brahe, a Dane. Brahe had become aware of Kepler's potential after reading a paper he had written while at university in Tübingen. After Kepler had been forced to leave his post as a mathematics lecturer at Graz in Austria, Brahe invited him to become his assistant in Prague under the patronage of Rudolph II, ruler of the Holy Roman Empire. Kepler took up the post in

Kepler one day 'saw a new light break' as he realised the orbits of the planets were ellipses

1600 and formulated a productive, if somewhat stormy, relationship with Brahe. One of the reasons the two argued was because Brahe rejected outright the Copernican view of the universe, which Kepler held in such high regard. The Dane had formulated his own alternative and rather obscure view on the rotation of the planets, which never caught on. Although history would subsequently prove Brahe to be wrong, his importance to Kepler, and astronomy in general, was that he was a brilliant observer of the skies and kept excellent records. When Brahe died in 1601, Kepler not only inherited his position of imperial mathematician in Rudolph's court but, crucially, his astronomical notes.

▸ THE ORBIT OF MARS

Using Brahe's records from the previous twenty years, Kepler set about trying to calculate and explain the orbit of Mars. Unfortunately, because he shared Copernicus's view that the planets orbited in perfect circles, the German struggled for the next eight years to produce a satisfactory conclusion. One day, he 'awoke from sleep and saw a new light break' as suddenly he realised that the planets did not rotate in perfect circles at all. They orbited around an ellipse, that is, a (flattened) circle with two 'centres' very close together. At a stroke, this would provide the simple mathematical explanation which had eluded Copernicus and **Ptolemy** when trying to predict the movement of the planets.

▸ KEPLER'S LAWS

In 1609, Kepler published his findings in *Astronomia Nova* or *New Astronomy*, which crystallised two 'laws' that would have a vital influence on our understanding of the universe. In a later book of 1619, *Harmonices Mundi* or *Harmonies of the World*, he added another important rule. These three together made up 'Kepler's Laws of Planetary Motion'. The first formalises his earlier discovery that the planets rotate in elliptical orbits with the sun at one of the centres, or focus points. The second states that all planets 'sweep' or cover equal areas in equal amounts of time regardless of which location of their orbit they are in. This is important because, as the sun is only one of two centres in a planet's orbit, a planet is nearer to the sun at some times than at others, yet it still 'sweeps out' the same area. What this means is that a planet must speed up when it is nearer the sun and slow down when it is further away. Kepler's third law finds that the 'period' (the time it takes to complete one full rotation – a year for the earth for instance) of a planet squared is the same as the distance from the planet to the sun cubed (in astronomical units). This allows distances of planets to be worked out from observing their cycles alone.

As well as providing the credible solution to predicting planetary motion that had previously proved so difficult, Kepler's findings would later act as the stimulus for questions that would lead to Isaac **Newton's** theory of gravity.

FURTHER ACHIEVEMENTS

Kepler's last major work was Tabulae Rudolphinae, Ruldolphine Tables *(1627) which were a painstakingly developed series of tables widely used in the next century to help calculate the positions of planets. He also made important discoveries in the field of optics,* *proposing the ray theory of light.*

He published a work of science fiction, A Dream, or Astronomy of the Moon *in 1634 At the time only two new stars visible to the naked eye had been discovered since antiquity. The second was observed by Kepler in 1604.*

WILLIAM HARVEY

1578–1657

CHRONOLOGY • 1609 Harvey appointed as physician at St Bartholomew's Hospital, London • 1618 Physician to James I • 1628 Publishes *On the Motion of the Heart and Blood in Animals* • 1651 Publishes *Essays on the Generation of Animals* • 1661 Marcello Malpighi uses his microscope to prove Harvey's assumptions regarding anastomoses

I f Johannes **Kepler** thrust astronomy into the modern world by 'completing' the work of Nicolaus **Copernicus** – who himself had confronted that of **Ptolemy** – then William Harvey was surely his anatomical equivalent. What **Galen** had begun and **Vesalius** had challenged, Harvey credibly launched into the modern arena with perhaps the most significant theory in his field of biology, before or since. What he postulated and convincingly proved was that blood circulated in the body via the heart – itself little more than a biological pump.

▶ A NEW THEORY

Galen had concluded that blood was made in the liver from food which acted as a kind of fuel which the body used up, thereby requiring more food to keep a constant supply. Vesalius, for all his corrections of Galen's work, added little to this theory. So it was left to the Englishman William Harvey, physician to King's James I and later Charles I, to prove his theory of circulation through rigorous and repeated experimentation on the 'royal' stock of animals over two decades. In the first instance, he had believed the heart could simply not produce the quantities of blood

To Harvey's mind, the blood was not used up, as Galen believed, but recycled around the body

required to support Galen's 'refuelling' theory. To Harvey's mind then, the only sound alternative was that blood was not used up but was recycled around the body. His dissections led him to correctly conclude that the arteries took blood from the heart to the extremities of the body, able to do so because of the heart's pump-like action. The veins, with their series of one-way valves, brought the blood back to the heart again. This rejected Galen's accepted explanation of how the body functioned.

Harvey published his findings in the 720-page *Exercitatio Anatomica de Motu Cordis et Sanguinis in Animalibus* or *Anatomical Exercise on the Motion of the Heart and Blood in Animals* at the Frankfurt Book Fair in 1628. He had, however, been lecturing on his theories of circulation since as early as 1616 but had taken a long time to publish his work. Rather like Copernicus, he was something of a perfectionist, partially explaining why he delayed for so long, but equally he feared a backlash against his theories for challenging Galen head on.

▶ DIVIDED OPINION

And rightly so. Although he initially received support from some academics, an equal number reacted with outrage and ridiculed his ideas. One of the areas where Harvey's work was weakest, which the author himself acknowledged but had been unable to solve, was that he struggled to offer a proven explanation for how the blood moved from the arteries to the veins. He speculated that the exchange took place through vessels too small for the human eye to see, which was confirmed shortly after his death with the discovery of capillaries by Marcello Malpighi with the recently-invented microscope. Harvey, though, had had no such luxury and even lost patients at his London practice as a result of the criticism directed towards him. By the time of his death, however, he had answered most of his detractors' objections and his conclusions became increasingly accepted, even before Malpighi's final proof.

▶ REPRODUCTION

In 1651, Harvey published another notable work, this time in the area of reproduction. *Exercitationes de Generatione Animalium* or *Essays on the Generation of Animals* included conjecture which rejected the 'spontaneous generation' theory of reproduction in mammals which had hitherto persisted. Instead, he suggested the only plausible explanation was that female mammals carried eggs which were somehow spurred into reproduction through interaction with the male's semen. While he did not foresee the egg itself being fertilised in the sense we now understand reproduction, his belief that the egg was at the root of all life was convincing, and gained acceptance long before the observational proof some two centuries later.

A MODERN METHODOLOGY

Harvey's significance comes not only from his discoveries, but also his methodology. As William Gilbert had begun in physics, and Francis Bacon had subsequently implored in all aspects of life, Harvey was the first to take a rational, modern, scientific approach to his observations in biology, sewing the seeds for a methodolgy that we can accept today. He cast aside the prejudices of his predecessors and only 'induced' conclusions based on the results of experiments which he could repeat identically again and again. It was a model which gained popularity following Harvey's success, and continues to be employed.

JOHANN VAN HELMONT

1579–1644

CHRONOLOGY • **1579** Van Helmont born to a wealthy, noble Brussels family
• **c. 1621** Put under house arrest by the Church • **1648** *Ortus
Medicinae* (*Origin of Medicine*), his collected papers, published by his son

While the sciences of physics, astronomy and anatomical biology were taking increasingly large strides towards the modern era, chemistry, as we know the subject today, was still lagging somewhat behind. Whether this was down to chance or because the other sciences had as their subjects more obviously observable phenomena, readily accessible for scrutiny, such as stars, motion or the human body, areas of chemistry were not immediately subject to the rational approach typical of the rest of the scientific revolution. Instead they had to wait for the posthumous publications of a slightly eccentric Flemish chemist, Jan Baptista van Helmont, for the transformation to begin.

▶ A KEEN OBSERVER

Born of a wealthy family, van Helmont had the luxury of being able to turn down paid employment and instead keep a rather solitary working life locked up in his own private laboratory conducting experiments. Van Helmont retained a belief in some areas of superstition such as the healing of wounds by treating the weapon that made them, although he insisted, contrary to the teaching of the Church, that this was an entirely

Van Helmont introduced the word 'gas', from a Flemish pronunciation of the Greek 'chaos'

natural phenomenon, and in no way miraculous. This attitude, predictably, soon led him into conflict with the Church, with the result that he spent much of his life under house arrest.

▶ THE PHILOSOPHER'S STONE

Another belief from the world of alchemy that Van Helmont retained was the existence of the Philosopher's Stone. Indeed, faith in this famous jewel's reality was one of the driving forces behind the 'science' of alchemy. To the alchemists the Stone was the elixir of life in solid, material form, and was capable, so the story went, of transmuting base metals into gold. Although imaginary, the pursuit of this fantastic gem has resulted in many important chemical discoveries.

Van Helmont still achieved enough to be considered by some to be the 'father' of modern biochemistry. In particular, he was the first to employ a calculated approach to his subject, most notably through the application of scientific measurement to the results of his experiments. He made use of a chemical balance and meticulously monitored his observations.

▶ WATER, WATER EVERYWHERE

For example, in one famous experiment, van Helmont planted and measured the growth of a willow tree over five years, during which time it gained 164 pounds. The scientist did this to 'prove' his belief that almost all matter was chiefly made up of water. While he rejected the **Aristotelian** belief in the 'four elements' (plus aether), van Helmont followed another early Greek scientist, Thales, in the fact that he was convinced of the dominance of water, even more so after his experiment. For five years the tree had only been fed on rainwater and the limited supply of soil in which it had been planted. Van Helmont found that the decrease in the soil's mass was only a few ounces, leading him to conclude that the tree was almost entirely composed of the water it had consumed. What he had failed to realise, ironically, given that he was the discoverer of the gas, was that about fifty per cent of the increased weight came from carbon dioxide in the air.

▶ SPLITTING AIR

Not that van Helmont failed to notice the existence of the gas in his other experiments, however. Indeed, the chemist's most important discovery was that gases other than 'air' existed at all. He realised that different elements gave off different gases when heated and was able to identify four distinct ones. He named them gas carbonum, gas sylvester of two variants, and gas pingue. These are the gases we now know as carbon dioxide, carbon monoxide, nitrous oxide and methane. In fact, van Helmont introduced to the world the term 'gas' itself, from the Flemish pronunciation of the Greek word 'chaos'.

FURTHER ACHIEVEMENTS

As well as chemistry, Van Helmont undertook studies in nutrition, digestion and physiology, applying the same scientific methodology to much of his work. He was, however, unable to divorce his fascination with the mystical from the majority of his studies, thereby discrediting much of their value when viewed from a modern perspective. More importantly, he recognised the law of the 'indestructibility of matter', realising, for instance, that metals dissolved in acid were only 'concealed' and could be regained in equal amounts. Van Helmont's son finally published the scientist's collected works in 1648 under the title Ortus Medicinae *or* Origin of Medicine.

RENÉ DESCARTES

1596–1650

CHRONOLOGY • **1596** Descartes born in La Haye, France • **1616** Graduated in law from the University of Poitiers • **10 November 1619** Descartes first begins to ponder the principles that would form his later work • **1637** *Discours de la Méthode* (*Discourse on Method*) published • **1637** *La Geométrie* (*Geometry*), published as an appendix to *Discours de la Méthode* • **1641** *Meditations on First Philosophy* published

René Descartes has been described as the first truly 'modern' mathematician and philosopher. Certainly his systematic, logical approach to knowledge was revolutionary, dominating philosophy for the next three centuries. Even more importantly, from the perspective of this book at least, it led to a new breakthrough which would greatly impact the future of mathematics and science.

Descartes initially gained a degree in law and spent a number of years in the military before eventually settling in Holland in 1628 where he composed all of his great works. In 1649 he accepted a post as personal tutor to Queen Christina of Sweden. A lifelong late riser and lover of a warm bed – where Descartes claimed to have undertaken his most profound thinking – he succumbed to the harsh Swedish weather. Within months he had contracted pneumonia and died.

▸ A REVELATION OF PHILOSOPHY

Three decades earlier, on the night of 10 November 1619, while campaigning with the

'Give me matter and motion and I will construct the universe'

army on the Danube, Descartes' life had changed forever when his influential journey began. He later claimed to have had a number of dreams on that date which formulated the principles behind his later work. In particular, it left him certain that he should pursue the theory that all knowledge could be gathered in a single, complete science and set about putting in place a system of thought by which this could be achieved. In turn, this left him to speculate on the source and truth of all existing knowledge. He began rejecting much of what was commonly accepted and vowed only to recognize facts which could be intuitively taken to be true beyond any doubt.

The full articulation of these processes came in Descartes' 1641 work *Meditations on First Philosophy*. The book is centred around his famous maxim 'Cogito, ergo sum' or 'I think, therefore I am,' from which he pursued all 'certainties' via a method of systematic, detailed mental analysis. This ultimately led him to a very detached, mechanistic interpretation of the natural world, reinforced in his 1644 metaphysical text the *Principia Philosophiae* or *Principles of Philosophy*, in which he attempted to explain the universe according to the single system of logical, mechanical laws he had earlier envisaged and, although largely inaccurate, would have an important influence even after **Newton's** more convincing explanations later in the century. He also regarded the human body as subject to the same mechanical laws as all matter, distinguished only by the mind which operated as a distinct, separate entity.

▸ MATHEMATICAL CERTAINTIES

Descartes passionately believed in the logical certainty of mathematics and felt the subject could be applied to give a superior interpretation of the universe. It is through this reasoning that his greatest legacy to mathematics and science came. In his 1637 appendix to the *Discourse*, entitled *La Geométrie*, Descartes sought to describe the application of mathematics to the plotting of a single point in space. This led him to the invention of what are now known as Cartesian Coordinates, the ability to plot a position according to x and y (that is, perpendicular) axes (and in a 3D environment by adding in a third 'depth' axis). Moreover, this method allowed geometric expressions such as curves to be written for the first time as algebraic equations (using the x, y and other elements from the graph).

The bringing together of geometry and algebra was a significant breakthrough and could, in theory at least, predict the future course of any object in space, given enough initial knowledge of its physical properties and movement. It is from his mathematical interpretation of the cosmos that Descartes would later claim, 'Give me matter and motion and I will construct the universe.'

THE 'COGITO'

Perhaps the most famous of philosophical maxims, 'Cogito, ergo sum', is best translated as 'I am thinking, therefore I am'. It was the result of a form of a thought experiment by Descartes, in which he resolved to cast doubt on any and all of his beliefs, in order to discover which he was logically justified in holding. He argued that although all his experience could be the product of deception by an evil demon (a more modern version has a brain in a vat, fed information by an evil scientist, an idea used in the film The Matrix*), the demon could not deceive him if he did not exist. That he can doubt his existence proves that he in fact exists.*

BLAISE PASCAL

1623–1662

CHRONOLOGY • **1642–44** Pascal invents and produces a calculating device to aid his father; in effect this is the first digital calculator • **1647** on the summit of the Puy de Dome he proves that air pressure decreases at higher altitudes • **1655** abandons his studies and enters the Jansenist retreat at Port-Royal • **1670** Pascal's Wager

Perhaps one of the lesser noted benefits of being a child prodigy is that if you die at an early age, you have still had sufficient time to fulfil your potential! Blaise Pascal, a Frenchman who passed away at just thirty-nine, was one such example. Although his time on earth was unfortunately cut short by poor health, and his contributions to mathematics and science severely limited by his abandonment of his studies in favour of religious devotion in 1655, he still had a significant influence within both fields of endeavour.

▸ MEASURING PRESSURE

During his twenties Pascal spent a large amount of time undertaking experiments in the field of physics. The most important of these involved air pressure. An Italian scientist, Evangelista Torricelli (1608–47), had argued that air pressure would decrease at higher altitudes. Pascal set out to prove this by using a mercury barometer. He took initial measurements in Paris and then, at the 1200m-high Puy de Dome in 1646, accompanied by his brother-in-law, he confirmed in no uncertain terms that Torricelli's speculation was true.

By the age of twenty-one, Pascal had finished what was effectively the first pocket calculator

▶ PASCAL'S LAW

More significantly, though, his studies in this area led him to develop Pascal's Principle or Law, which states that pressure applied to liquid in an enclosed space distributes equally in all directions. This became the basic principle from which all hydraulic systems derived, such as those involved in the manufacture of car brakes, as well as explaining how small devices such as the car jack are able to raise a vehicle. This is because the small force created by moving the jacking handle in a sizeable sweep equates to a large amount of pressure sufficient to move the jack head a few centimetres. Applying the lessons of his studies in a practical way, Pascal went on to invent the syringe and, in 1650, the hydraulic press.

▶ CHILD PRODIGY

In spite of these developments, however, Pascal is probably better remembered for his work in the area of mathematics. It was here that he showed his genius from an early age. For example, having independently discovered a number of Euclid's theorems for himself by the age of eleven, he went on to master *The Elements*, the great mathematician's definitive text, by twelve. When he was sixteen he published mathematical papers which his older contemporary **Descartes** at first refused to believe could have been written by one so young. In 1642, still only nineteen,

Pascal began work on inventing a mechanical calculating machine which could add and subtract. He had finished what was effectively the first digital calculator by 1644 and presented it to his father to help him in his business affairs.

▶ THEORY OF PROBABILITY

It was not until later in his short life, around 1654, that Pascal jointly made the mathematical discovery which would have the most impact on future generations. It had begun with a request by an obsessive gambler, the Chevalier de Méré, for assistance in calculating the chance of success in the games he played. Together with Pierre de Fermat, another French mathematician, Pascal developed the theory of probabilities, using his now famous Pascal's Triangle, in the process. As well as its obvious impact upon all parts of the gambling industry, the importance of understanding probability has had subsequent application in areas stretching from statistics to theoretical physics.

The SI unit of pressure – the pascal – and the computer language, Pascal (named in honour of his contribution to computing through his invention of the early calculator), are named after him in recognition of two of his main areas of scientific success.

Seven of the calculating devices that he produced in 1649 survive to this day.

PASCAL'S WAGER

Like many of his contemporaries, Pascal did not separate his science from philosophy, and in his book Pensees, he applies his mathematical probability theory to the perennial philosophical problem of the existence of God. In the absence of evidence for or against God's existence, says Pascal, the wise man will choose to believe, since if he is correct he will gain his reward, and if he is incorrect he stands to lose nothing, an interesting, if somewhat cynical argument.

ROBERT BOYLE

1627–1691

CHRONOLOGY •**1638** Boyle takes up studies in Geneva, after four years at Eton College
• **1644** Retires to his estate in Dorset, to pursue his own studies • **1654**
moves to Oxford, where he meets and befriends Robert Hooke • **1659** Boyle and Hooke carry out
experiments with vaccuum • **1662** Formulates his famous chemical dictum, Boyle's Law

Born in the modern day Republic of Ireland as the fourteenth child of the richest man in what was then part of Great Britain, Boyle enjoyed all the privileges of an aristocratic education. Schooled at Eton and then privately, he continued his studies as he undertook a long European tour from 1639 to 1644. Eventually he returned to an inherited estate in Dorset, England, largely avoiding the worst excesses of the Civil War. Here he began his scientific studies. In 1656 he moved to Oxford where, with the philosopher John Locke, and the architect Christopher Wren,

he formed the Experimental Philosophy Club. He also met Robert **Hooke**, who became his assistant and the two formed a productive partnership. It was together with Hooke that Boyle began making the discoveries for which he became famous.

▶ BOYLE'S LAW

Chief amongst these was the expression of what is now known as Boyle's Law (also independently discovered by the French scientist Edme Mariotte) which established a direct relationship between air pressure and volumes of gas. By using

Boyle's Law: pressure is inversely proportional to volume at constant temperature

mercury to trap some air in the short end of a 'J' shaped test tube, Boyle was able to observe the effect on its volume by adding more mercury. What he found was this: if he doubled the mass of mercury (in effect, doubling the pressure), the volume of the air in the end halved; if he tripled it, the volume of air reduced to a third, and so on. As long as the mass and temperature of the gas were constant, his law concluded that the pressure and volume were inversely proportional.

▸ THE VACCUUM PUMP

This experiment was the culmination of a series of other tests involving air and its effects. They had begun shortly after Boyle had moved to Oxford and progressed rapidly when Robert Hooke had constructed an air pump upon his request. The pump was able to create the best man-made vacuum to date and through experiments involving bells, animals and candles, Boyle was able to draw a number of important conclusions. He found that sound could not travel through a vacuum and required air in order to do so. Air was required for respiration and combustion, and not all of the air was used up during breathing and burning processes. In addition, he proved Galileo's proposal that all matter fell at equal speed in a vacuum.

▸ THE SCEPTICAL CHEMIST

In 1661, Boyle published *The Sceptical Chemist* which criticised the Aristotelian view of a universe composed of only four elements (earth, water, air and fire), plus aether in wider space. The text helped pave the way to our current view of the elements. Although he did not describe elements exactly as we understand them today, he believed that matter consisted at root of 'primitive and simple, or perfectly unmingled bodies' which could combine with other elements to form an infinite number of compounds. This was an extension of his support for early atomic theory, believing in what he described as tiny 'corpuscles'. In spite of an interpretation which does not entirely correspond with the modern view, his importance was in promoting an area of thought which would influence the later breakthroughs of Antoine Lavoisier (1743–93) and Joseph Priestley (1733–1804) in the development of theories related to chemical elements.

THE INFLUENCE OF ROBERT BOYLE

Robert Boyle made a number of contributions to the history of science, but perhaps most significant is his claim to being the man responsible for the establishment of chemistry as a distinct scientific subject in its own right. Like his idol Francis Bacon, he experimented relentlessly, accepting nothing to be true unless he had firm empirical grounds from which to draw his conclusions.

Boyle's other important legacies were the creation of flame tests in the detection of metals, as well as tests for identifying acidity and alkalinity.

Boyle was also a founder member of the Royal Society, the longest running scientific society in the world. But it was his insistence on publishing chemical theories supported by accurate experimental evidence – including, for the first time, details of apparatus and methods used as well as failed experiments – which would have the most impact upon modern chemistry.

CHRISTIAAN HUYGENS

1629–1695

CHRONOLOGY • **1655** Huygens discovers Titan, Saturn's largest moon • **1657** Clock constructed according to his groundbreaking design • **1658** *Horologium* (*The Clock*) published • **1673** *Horologium Oscillatorium* (*The Clock Pendulum*) published • **1690** *Traité de la Lumière* (*Treatise on Light*) published

The Dutchman Christiaan Huygens is widely considered to be the second most important physical scientist of the seventeenth century. Son of the distinguished diplomat and scholar Constantjin Huygens, Christiaan was acquainted from an early age with notables such as René Descartes, a friend of the family.

Unfortunately for him, one of his key propositions on the behaviour of light contrasted directly with the first and most important, originally proposed by Isaac Newton. As a result Huygens' theory was largely ignored for over a

century. His other achievements in time measurement did impact immediately, however, helping his science to progress in a way it otherwise would have struggled to do.

▶ AGAINST NEWTON

Isaac Newton articulated a particle theory of light, believing it to be made up of 'corpuscles'. This was a view he summarised in his 1704 text *Opticks* but had held for the preceding decades. He vigorously challenged anyone who tried to contradict this opinion, as both Leibniz and Robert Hooke (1635–1703) – who shared similar

His key idea on the behaviour of light went against Newton: it was ignored for a century

views to the Dutchman – were to find out. Huygens believed light actually behaved in a wave-like fashion, in a method which became known as the 'Huygens Construction,' which he outlined in his 1690 work *Treatise on Light* (although he had first expressed it in 1678). This opinion much more satisfactorily explained the way light reflected and refracted, and correctly anticipated that in a denser medium light would travel more slowly. Although the modern interpretation is that light can behave in both a particle and wave-like fashion depending on the situation, Huygens's view, when rediscovered and championed by Englishman Thomas Young (1773–1829), in the early nineteenth century, would eventually become the more commonly accepted version. Such was the dominance of Isaac Newton, however, that Huygens's theories were totally ignored for the whole of the eighteenth century, and they still faced fierce resistance in Young's time.

▸ THE PENDULUM CLOCK

Of much more immediate impact, though, were Huygens's breakthroughs in clock-making. Ever since the time of **Galileo** (1564–1642), scientists had been aware that a swinging pendulum could keep a regular beat and they had hoped to use this knowledge to create an accurate time-measuring device. They had been unsuccessful.

Huygens realised that this was partly because a pendulum mimicking a circle's curve did not maintain a perfectly equal swing and in order to do this it actually needed to follow a 'cycloidal' arc. This discovery set him on the path to designing the first successful pendulum clock. He had it constructed in 1657 and announced his creation to the world in his 1658 book *Horologium* or *The Clock*. The invention was of monumental importance to the progress of physics, for without an accurate method of measuring time the progress of the subject over the following centuries would have been severely hampered.

Huygens backed up his practical findings with mathematical explanations describing a pendulum's swing in his 1673 work *Horologium Oscillatorium* or *The Clock Pendulum*. The text included a number of other dynamic explanations and anticipated the first of Newton's motion laws, the 'law of inertia', which states: an object moving in a straight line will continue to move in the same way indefinitely until it meets another force.

Huygens was an associate of Leibniz, whom he supported during his controversial bout with Isaac Newton over the law of gravity. Despite this, and despite Huygens' opinion that Newton's theory of gravity was incomplete without a mechanical explanation, as expressed in the *Principia*, Newton was a staunch admirer of the Dutchman.

ASTRONOMY AND LIGHT

As well as being an accomplished physicist, Christiian Huygens was also a keen astronomer and made some important contributions in this area. He devised a much-improved telescope and used it to make a number of findings, including the discovery of Saturn's biggest moon, Titan, in 1655. In addition, he observed and accurately explained Saturn's ring system.

Huygens' hypothesis that light is a wave was largely ignored at the time as it conflicted with Newton's theory which proposed that light had a particle structure. Both were in fact correct.

Huygens was one of the founding fathers of the French Academy of Sciences in 1666, and was granted a larger pension from that body than anyone else.

ANTON VAN LEEUWENHOEK

1632–1723

CHRONOLOGY • **1673** Van Leeuwenhoek begins correspondence with the Royal Society • **1674** First to observe protozoa • **1677** First to observe human spermatozoa • **1683** First to observe bacteria • **1684** First to observe red blood cells

Who said you had to be a full time scientist with money and an aristocratic background to make world-changing discoveries? Probably the vast majority of people who lived in the seventeenth century, when most scientists came either from the nobility – possessing the independent wealth to undertake research with no need for a job – or were funded by them through patronage. Not quite so for Anton van Leeuwenhoek, a humble Dutch draper, who, despite little formal education, went on to entertain kings and queens with his remarkable revelations.

▶ HIS HOBBY

Born and remaining all his life in Delft in the Netherlands, van Leeuwenhoek became an apprentice linen-draper at the age of 16 and went on to open his own business in the town around 1654. In 1660 he took on a better-paid position in the town's law courts. This gave him greater means and more spare time to pursue the subject he would bring to impact on history: van Leeuwenhoek had developed a passion for microscopy, and by 1660 was devoting all the spare time he could get to producing lenses with a greater magnification than had ever been made before.

He discovered that when his faeces were 'a bit looser than usual', protozoa were observed

▶ **THROUGH THE LOOKING GLASS**

Van Leeuwenhoek kept secret his methods for producing the lenses during his entire ninety years. Even though his finest single, short focal length lenses could enlarge a specimen by up to three hundred times, it is believed he employed an additional technique, perhaps some form of illumination, to view the miniscule 'animalcules' he observed. Another major discovery was protozoa, effectively tiny one-celled plants, which he came across in water specimens in 1674. In more recent scientific studies protozoa would be linked to a number of tropical diseases including, most significantly, malaria and amoebic dysentery. In 1683, and perhaps even more importantly, van Leeuwenhoek observed bacteria for the first time. They were smaller than protozoa and were later linked to diseases such as cholera and tetanus, as well as their treatment.

▶ **ANIMAL REPRODUCTION**

In between these findings, van Leeuwenhoek discovered spermatozoa. The story goes that in 1677 his contemporary Stephen Hamm brought the microscopist a sample of human semen. Upon examination he discovered the short-lived sperm, reinforcing his opinion of their importance in reproduction by finding similar creatures in the semen of frogs, insects and other animals. He made detailed and exact observations of both fleas and ants, proving that the former were generated from eggs like any other insects, rather than arising spontaneously. He also showed that the eggs and pupae of ants were phenomena occurring at two entirely different stages. From this he correctly claimed their existence as evidence of his belief that the commonly held view of 'spontaneous generation' of insects and other small organisms was wrong.

Other important discoveries included the observation of red blood cells in 1684, providing further support for Marcello Malpighi's 1660 work on blood capillaries, which in itself had been so important in reinforcing William **Harvey's** speculation on the transfer of blood from the arteries to the veins. From ants to shellfish, van Leeuwenhoek also undertook a range of further studies, including observations of animal life .

▶ **LANGUAGE BARRIER**

In keeping with his lack of academic training, van Leeuwenhoek wrote up his findings in Dutch rather than the scholarly Latin, and published little directly. Instead he was introduced to the Royal Society of England via correspondence in 1673. For the rest of his life, van Leeuwenhoek would subsequently write regularly to the Royal Society in Dutch, outlining his latest discoveries, for he knew no English. In all, the society translated and printed some 375 entries in their publication *Philosophical Transactions* before van Leeuwenhoek's death.

FURTHER ACHIEVEMENTS

Van Leeuwenhoek's letters, and his subsequently assembled collected works, made the part-time scientist world famous and brought many noble visitors to Delft. Amongst those who came to see the animalcules first hand were James II of England and Peter the Great of Russia.

When van Leeuwenhoek died he left behind

247 complete microscopes, nine of which survive to this day. One of his microscopes had a resolution of 2 micrometers.

Examining his own faeces, he observed that 'when of ordinary thickness' there were no protozoa observed, but when 'a bit looser than ordinary' protozoa were observed.

ROBERT HOOKE

1635–1703

CHRONOLOGY • **1656** Hooke meets Robert Boyle at Oxford • **1659** Devises the pump, the most efficient creator of a vacuum at the time • **1662** Becomes the first Curator of Experiments to the Royal Society • **1665** *Micrographia* (*Small Drawings*) published • **1670** Discovers the law of elasticity

Perhaps one of the most 'underrated' scientists of the seventeenth century, Robert Hooke, an Englishman, experimented and made advances in a wide range of scientific areas. Yet because of this breadth of coverage, he seldom developed any of his concepts to their fullest extent. This explains why he rarely gained credit for them. Indeed, it is arguable that his role as a provider of ideas to others is his most important legacy.

▶ BOYLE'S ASSISTANT

The most obvious example of his contributions to

others was the work he undertook with Robert Boyle at Oxford, where they met in 1656. Boyle, as the aristocrat, was clearly the dominant partner in the relationship, in social terms at least. Hooke, as his assistant, acted on Boyle's instructions, yet many of his creations were worthy inventions in their own right. The most obvious example is the air pump that he devised in 1659, the most efficient vacuum creator of its time. It enabled Boyle to go on to make many of his discoveries.

▶ PROVIDER OF IDEAS

Moreover, Boyle was responsible, albeit indirectly,

Hooke accused Newton of plagiarism, sparking a bitter lifelong feud between the two

for keeping Hooke in his position as jack of all sciences, master of none. The aristocrat had been influential in having Hooke elevated to the position of Curator of Experiments for the Royal Society in 1662. While the prestige of the role pleased Hooke, the job requirement of showing 'three or four considerable experiments' to the Society at each of its weekly meetings was almost certainly the factor that ensured Hooke would never have the time to develop any of his findings fully.

▶ A SOURCE OF IDEAS

Another scientist to whom Hooke felt he had provided source material was the Dutch physicist Christian **Huygens**. Huygens is credited with creating the influential wave theory of light, which he published in 1690. Yet as early as 1672, Hooke had explained his discovery of diffraction (the bending of light rays) by suggesting that light might behave in a wave-like fashion.

Isaac **Newton** vehemently argued against Hooke's theory of light, beginning a bitter feud which would continue for the rest of Hooke's life. Hooke also claimed to have discovered one of the most important theories credited to Newton, arguing that the latter had plagiarised his ideas from correspondence between the two during 1680. Certainly, Hooke's letters suggested some notion of universal gravitation and hinted at an understanding of what later became Newton's

law of gravity. In spite of this, though, it is unquestionable that Newton's mathematical calculations and endeavours in proving the law give him a much stronger claim.

Hooke's countless experiments did, however, result in some other discoveries solely credited to him. He was, for example, the first to describe the universal law that all matter will expand upon heating. He is credited with the law of elasticity, discovered in 1670. Also known as Hooke's Law, it states that the strain, or change in size, placed upon a solid – when stretched – is directly proportional to the stress, or force, applied to it. Hooke was also the first person to use the word 'cell' in the scientific sense understood by us today, after observing the properties of cork under one of the powerful microscopes that he had developed. This word was used in his 1665 work *Micrographia* or *Small Drawings,* which also included many other advances such as Hooke's theory of combustion, as well as other discoveries of the microscope. These included crystalline structure of snow, and studies of fossils which led to the proposition that they were the remains of once living creatures. He suggested that whole species had lived and died out long before man, centuries before Charles Darwin came to the same conclusion.

Hooke also made discoveries in astronomy, locating Jupiter's Great Red Spot, and proposed that the huge planet rotated on its axis.

FURTHER ACHIEVEMENTS

Hooke's inventions were greatly influential. He either invented or significantly improved the reflecting telescope, compound microscope, dial barometer, anemometer, hygrometer, balance spring (for use in watches), quadrant, universal joint and iris diaphragm (later used in cameras). He also showed impressive vision,

foreseeing the development of the steam engine and the telegraph system.

The inventions of the compound microscope and a balance spring for use in watches are also credited to him. Beyond this he was an accomplished architect who designed parts of London following the great fire of 1666.

SIR ISAAC NEWTON

1642–1727

CHRONOLOGY • **1670–71** Newton composes *Methodis Fluxionum* (*Method of Fluxions*), his main work on calculus which is not published until 1736 • **1672** New *Theory about Light and Colours* published. It is his first published work • **1687** *Philosophiae Naturalis Principia Mathematica* (*Mathematical Principles of Natural Philosophy*), known as the *Principia*, published • **1704** *Opticks* published

So many extensive books and articles have been written on the life and impact of Sir Isaac Newton over the last three centuries it is impossible to do his achievements justice in a short entry like this. He is quite simply one of the greatest scientists of all time.

▸ A SLOW BEGINNING

His early years did not necessarily suggest, however, he would end up as such. Born and brought up in the quiet village of Woolsthorpe in

Lincolnshire, England, and schooled in the nearby town of Grantham, he was not particularly noted for academic achievements as a child. Even on entry to Trinity College, Cambridge, he did not stand out until, ironically, the University was forced to close during 1665 and 1666 due to the high risk of plague. Newton returned to Woolsthorpe and began two years of remarkable contemplation on the laws of nature and mathematics which would transform the history of human knowledge. Although he published nothing during this period, he formulated and tested

'If I saw further than others, it is because I was standing on the shoulders of giants'

many of the scientific principles which would become the basis for his future achievements.

However, it would often be decades before he returned to his earlier discoveries. For example, his ideas on universal gravitation did not re-emerge until he began a controversial correspondence on the subject with Robert **Hooke** in around 1680. Furthermore, it was not until Edmond **Halley** challenged Newton in 1684 to find out how planets could have the elliptical orbits described by Johannes **Kepler**, and Newton replied he already knew, that he fully articulated his law of gravitation. Yet he had begun work on the subject back in the 1660s in Woolsthorpe after famously seeing an apple fall from a tree and wondering if the force which propelled it towards the earth could be applied elsewhere in the universe. After his declaration to Halley, Newton was forced to recalculate his proof, having lost his original jottings, and the result was published in Newton's most famous work *Philosophiae Naturalis Principia Mathematica* (1687). This law of gravitation proposed that all matter attracts other matter with a force related to the combination of their masses, but this attraction is weakened with distance, indeed, in inverse proportion to the square of their distances apart. This universal principle applied just as equally to the relationship between two small particles on earth as it did between the sun and the planets, and Newton was able to use it to explain Kepler's elliptical orbits.

▸ NEWTON'S LAWS OF MOTION

In the same work, Newton built on earlier observations made by **Galileo** and expressed three laws of motion which have been at the heart of modern physics ever since. The 'law of inertia', states that an object at rest or in motion in a straight line at a constant speed will carry on in the same state until it meets another force. The second stated that a force could change the motion of an object according to the product of its mass and it acceleration, vital in understanding dynamics. The third declares that the force or action with which an object meets another object is met by an equal force or reaction.

Aside from the wide ranging uses for the laws Newton outlined in the *Principia*, the important point is that all historical speculation of different mechanical principles for the earth from the rest of the cosmos were cast aside in favour of a single, universal system. It was clear that simple mathematical laws could explain a huge range of seemingly disconnected physical facts, providing science with the straightforward explanations it had been seeking since the time of the ancients. Newton's insistence on the use of mathematical expression of physical occurrences also underlined the standard for modern physics to follow.

FURTHER ACHIEVEMENTS

Newton achieved major breakthroughs in other areas too. His proof that white light was made up of all the colours of the spectrum was outlined in his 1672 work New Theory about Light and Colours. *In* Opticks *(1704), he also articulated his influential (if partially innacurate) particle or corpuscle theory of light. Another achievement significant to mathemat-* *ics was his invention of the 'binomial theorem'.*

Newton had a practical side too, inventing the reflecting telescope in the 1660s. This new instrument bypassed the focusing problems caused by chromatic aberration in the refracting telescope of the type Galileo had created.

During his time as master of the Mint twenty-seven counterfeiters were executed.

EDMUND HALLEY

1656–1742

I t is frequently the case with the famous dead that they are remembered for a single discovery, action, theory or invention. Edmund Halley, the English astronomer and mathematician, is perhaps the greatest example of this phenomenon, renowned today for the discovery of the comet which bears his name. Yet more than almost any other scientist featuring in this book, his academic interests were broad and wide, with an impact far greater than the observation of a single cosmological boomerang.

▸ HALLEY'S COMET

Not that his most celebrated achievement should be underestimated. Halley's Comet, as it became known for the first time when it reappeared sixteen years after the astronomer's death in 1758, exactly when he said it would, was the first comet whose return had been predicted. Halley had come to this conclusion after observing the body for himself in 1682. After further research he deduced that other comets, visible in 1531 and 1607, were so similar in characteristic to the one that he had seen that they were in fact probably a

Halley should be remembered for more than the observation of a single cosmic boomerang

single visitant simply returning at an interval of seventy-six years. The findings, which also calculated the orbits of twenty-three other comets, and published in 1705 as *A Synopsis of the Astronomy of Comets*, were seminal in the subsequent approach to the study of the subject.

▸ SOUTHERN SKIES

Halley's astronomical interests were not restricted to comets, however, and he contributed many other important studies. In 1718, he demonstrated that stars must have a 'proper' motion of their own by making comparisons between **Ptolemy's** catalogue and the position of the stars in his own time. He also observed the moon's full nineteen-year cycle and after doing so confirmed the theory of secular acceleration that he had originally predicted in 1695. In 1716, he proposed a way of calculating the Earth's distance from the Sun from transits of the planet Venus across the Sun's disc. One of his greatest celestial achievements was also one of his first. At the age of 20 he travelled on an East India Company ship to St Helena in order to map the stars in the southern hemisphere. He left Oxford University without completing his degree in order to do it. After two years of study on the remote island his publication of *The Catalogue of Southern Stars* in 1679 was not only the first accurate mapping of the southern skies, but also

the first telescopically determined survey of the stars that he observed.

▸ NOT JUST ASTRONOMY

Closer to home, Halley was to gain credit in many other arenas. He is considered by some to be the founder of geophysics, beginning with his publication of a map in 1686 of the Earth's prevailing winds, and went on to prepare detailed maps of the tides and magnetic variation. He undertook work on the salinity and evaporation of lakes between 1687 and 1694, using his results to offer theories on the Earth's age. Halley developed a mathematical law which demonstrated the relationship of height to air pressure, allowing him to go on to make improvements in the design of the barometer. The population of the city of Breslau was the subject of the mortality tables he published in 1693, pioneering work in social statistics which later influenced the life insurance industry. The size of the atom, the optics of the rainbow and even the design of the diving bell did not escape the scrutiny of the man. Halley was definitely not just an astronomer.

As well as commanding the *Paramour*, a Royal Navy man of war, from 1698 to 1700, he was also a prolific mapmaker, showing prevailing winds, tides and magnetic variations in his cartography.

Halley's Comet will return to the skies in 2062.

HALLEY AND NEWTON

Despite his many achievements, it is arguable that perhaps the most important way Halley can be seen to have changed the world is in his friendship with Newton. He met him for the first time at Cambridge in 1684 and from then on would have an important role in the development and presentation of the theory of gravitation. He encouraged Newton to undertake his greatest work, the Principia, in the first

place. He went on to edit and proof-read the text, write the preface and perhaps most importantly of all, to finance its publication himself in 1687, when the Royal Society failed to do so.

Had Edmund Halley not been born, his comet would still exist, albeit under a different name. Newton's Principia, at least in the form the world knows it today, would almost certainly not.

THOMAS NEWCOMEN

1663–1729

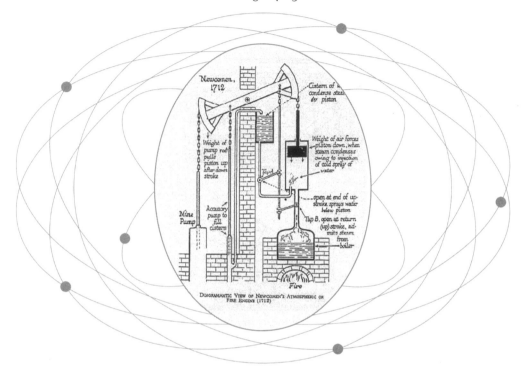

Newcomen, 1712

Cistern of condense steam or piston

Weight of air forces piston down, when steam condenses owing to injection of cold spray of water

Weight of pump rod pulls piston up after down stroke

open at end of up-stroke sprays water below piston

Tap B, open at return (up) stroke, admits steam from boiler

Mine Pump

Accessory pump to fill cistern

Fire

DIAGRAMMATIC VIEW OF NEWCOMEN'S ATMOSPHERIC OR FIRE ENGINE (1712)

CHRONOLOGY • 1663 Newcomen born in Dartmouth, Devon, England • 1705 Begins working to build a steam engine • 1712 Successfully builds and puts into use his improved steam engine • 1729 Dies in London

I f the Industrial Revolution changed the world, then the man who made available the power source which facilitated the transformation must be heralded. This man was Thomas Newcomen, the English inventor of the world's first commercially successful low-pressure steam engine.

Of course, Newcomen had not set out to alter the development of society quite so dramatically. He had begun life as a simple ironmonger and smith in his hometown of Dartmouth in Devon, establishing a living in his chosen trade long before he began work on his ground-breaking invention. It was, however, his occupation which

brought to his attention the problems which inspired his invention.

▶ THE PROBLEM

Many of Newcomen's most important customers owned mines. They described to him the problem they encountered when forced to dig deeper mines to meet the growing demand for natural resources such as coal, tin and iron ore. Owing to the depth, they were increasingly hindered by flooding. The solution was to pump this water out of the mines but with only horse or manpower available to do the job, it was an expensive and slow task.

Thomas Newcomen provided the power sources for the Industrial Revolution

▸ FAILED ATTEMPTS

The idea of using atmospheric pressure as a new kind of power source to be employed in carrying out repetitive, mechanical work like pumping had been known to engineers before Newcomen. It had been proved that when a vacuum was created, air, when given the opportunity, would rush into it with considerable force. But nobody had ever harnessed this discovery successfully into a practical power supply. In 1698, an English engineer called Thomas Savery (1650–1715) had made an attempt at it through his design and patent of the 'Miner's Friend', a high-pressure steam pump engine. Due to technological and practical limitations, however, it was never successfully employed for pumping.

▸ THE SOLUTION

It was against this context that Newcomen decided to begin work in 1705 on building a steam engine to take advantage of atmospheric pressure. By 1712, he had solved the problem and his engine was successfully constructed and used for pumping in South Staffordshire Colliery. The design involved heating water underneath a large piston which was encased in a cylinder. Steam that was released as a result of the heating forced the piston upwards. A jet of water was then released from a tank above the piston. The sudden cooling of the steam made it condense, creating a partial vacuum which atmospheric pressure then pushed down on, forcing the piston downwards again. The piston was attached to a two-headed lever, the other side of which was attached to a pump in the mineshaft. As it moved up and down, the lever moved likewise and a pumping motion was created in the shaft which could be used to eject flood water. The first engine could remove about 120 gallons per minute, completing about twelve strokes in that time, and had the equivalent of about 5.5 horsepower.

Even thought the engine was still not particularly powerful, was hugely inefficient to run, and burnt large amounts of coal, it would work reliably twenty-four hours a day and was far better than the previous alternatives. Consequently, even though each one cost an expensive £1000 to build, they were highly successful commercially and as a result more than a hundred were installed, chiefly in Britain's mines and factories, before Newcomen's death in 1829. Sales continued to increase across Britain and Europe for the next one hundred years. Even though more efficient engines were to follow, such as that invented by James Watt (1736–1819), its relative simplicity, reliability – and lower price tag than the competition – ensured that working engines continued to be used well into the twentieth century. By then, the Industrial Revolution had changed the world, and Newcomen's harnessing of steam and atmospheric pressure had been at its helm.

A FORGOTTEN GENIUS

The steam engine, although originally developed by Newcomen for use in mines, went on to become one of the cornerstones of the Industrial Revolution. It was quickly developed by engineers like James Watt and Richard Trevithick into the steam locomotive, eventually going on to power the ocean-going ironclads which cut sailing times to and·from Britain so drastically. Today, the credit for the steam engine is usually given to James Watt while the name Thomas Newcomen remains shrouded in obscurity. And although he undoubtedly changed the world, there is no single known portrait of Newcomen in existence.

DANIEL FAHRENHEIT

1686–1736

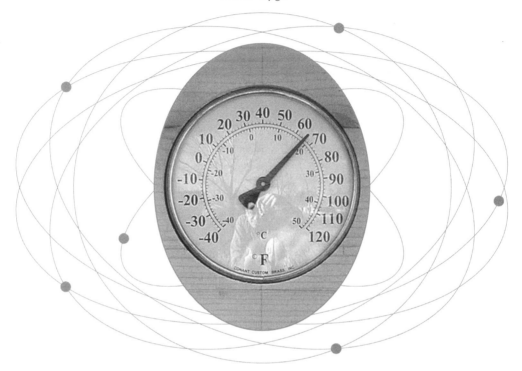

CHRONOLOGY • **c. 1701** Fahrenheit arrives in Amsterdam, Holland, apprenticed to merchants • **1709** Develops a superior alcohol thermometer • **1714** Invents the first mercury thermometer • **1715** Develops the Fahrenheit temperature scale

Daniel Fahrenheit spent most of his working life in the Netherlands. He was born in the Polish city of Danzig, now Gdansk, the oldest of five children. At the age of fifteen his parents died after eating poisonous toadstools and while his four siblings were sent to foster homes he became apprenticed to merchants who sent him to Amsterdam. Here he became interested in thermometry, particularly the primitive thermometers invented in Florence around 1640. Fahrenheit borrowed money against his inheritance to develop the idea and left his apprenticeship – this forced him to go on the run.

The measurement and description of temperature is so commonplace today that it is virtually impossible to imagine a world without it. Yet as late as the start of the eighteenth century, scientists were still struggling to find a reliable device which would accurately measure temperature, and a uniform scale by which to describe the limited measurements they could make.

▶ PRIMITIVE THERMOMETERS

Galileo (1564–1642) had, in fact, been the first to create a primitive kind of thermometer. He had used his knowledge that air expanded when heated and contracted when cooled to build his

The Fahrenheit scale, perhaps unsurprisingly, is named for its discoverer, Daniel Fahrenheit

instrument. By placing a cylinder tube in water, he noticed that when the air was hotter it pushed the level of the water in the device downwards, just as it rose when the air cooled. He soon realised the reading was unreliable, however, because the volume, and therefore the behaviour, of the air also fluctuated according to atmospheric pressure. Gradually, scientists began using other, more stable substances to improve the accuracy of the reading, with alcohol being introduced as a possible substitute later in the century.

▸ THE MERCURY THERMOMETER

Fahrenheit eventually made the thermometer reliable and accurate enough for the purposes scientists required. A producer of meteorological instruments, he first achieved progress in 1709 with the development of an alcohol thermometer far superior to any that had previously been created. However, it was in building upon Guillaume Amontons' (1663–1705) work on the properties of mercury that Fahrenheit truly took the measurement of temperature into another domain. He invented the first successful mercury thermometer in 1714, particularly useful in its application across a wide range of temperatures.

▸ THE FAHRENHEIT SCALE

In 1715 he complemented his breakthroughs in instrument-making with the development of the now famous Fahrenheit temperature scale. Taking zero degrees Fahrenheit (0°F) to be the lowest temperature he could produce (from a blend of ice and salt), he used the freezing point of water and the temperature of the human body as his other key markers in its formulation. In his initial calculations this placed water's freezing point at 30°F and the body's at 90°F. Later revisions changed this to the now well-known 32°F for water and 96°F for humans. The boiling point of water worked out to be 212°F, meaning that there were a hundred and eighty incremental steps between freezing and boiling. The scale became widely used, particularly in English speaking countries where it largely remained dominant until the 1970s. The fahrenheit scale is still in common use throughout the USA.

▸ FAHRENHEIT'S SUCCESSORS

Across much of the scientific community Fahrenheit's range has been superseded by the Celsius scale. This metric scale, with water freezing at 0°C and boiling at 100°C, was begun by the Swede Anders Celsius (1701–44), and revised by his countryman Carolus Linnaeus (1707–78) into the scale we know today (Celsius originally had the scale the other way round, with boiling point at 0°C and freezing at 100°C!). The Celsius scale is also known as the Centigrade scale, from the Latin meaning 'one hundred steps'. A conversion can be made from degrees Fahrenheit to degrees Celsius by subtracting thirty-two and multiplying this figure by five, then dividing by nine.

THE INFLUENCE OF FAHRENHEIT

The Kelvin scale is more suitable for scientific purposes and the Celsius scale is neater, based as it is on decimals. The advantage of using the Fahrenheit scale is that it is designed with everyday use in mind, rarely needing, for example, negative degrees. Fahrenheit, using his glass-blowing skills to create his thermometer, found that the boiling points of different liquids varied according to fluctuations in atmospheric pressure; the lower the pressure, the lower the boiling point of water, a useful fact for anyone wishing to make tea at high altitude!

BENJAMIN FRANKLIN

1706–1790

Benjamin Franklin had a rare genius. Unlike most entrants in this book, whose outstanding talents are generally restricted to the scientific, the American Franklin was brilliant in a wide range of arenas. In a five-year period between 1747 and 1752, he contributed more to science than most scientists would achieve in a lifetime of dedicated study. Yet, during other periods of his life, he operated in, and conquered, completely different fields. He was a master printer and publisher, a successful journalist and satirist, an inventor, a world famous ambassador

and, probably most notably of all, a politician at a vital time in American history. Indeed, Franklin was one of the five signatories of the Declaration of Independence from Great Britain in 1776 and was a key participant in the later drafting of the American Constitution.

▸ STUDYING ELECTRICITY

Yet Franklin merits an entry in this book for his achievements in physics alone – he was a pioneer in understanding the properties and potential benefits of electricity. Although the phenomenon of electricity had been noted since the time of the

Franklin's legacy, in addition to his many inventions, was essentially one of learning

ancients, very little was known about it from a scientific perspective, and many considered the extent of its usefulness to be limited to 'magic' tricks. At around the age of forty, however, Franklin became fascinated by electricity and began to experiment with it, quickly realising it was a subject worthy of scientific study and research in its own right. So, he sold his printing interests and dedicated himself for the next five years to understanding it.

▸ FLYING A KITE

Although Franklin wrongly believed electricity was a single 'fluid' (this was in itself an advance on earlier theories which posited the idea of two different fluids), he perceived this fluid to somehow consist of moving particles, now understood to be electrons. More importantly, he undertook important studies involving electrical charge and introduced the terms 'positive' and 'negative' in explaining the way substances could be attracted to or repelled by each other according to the nature of their charge. He also believed these charges ultimately cancelled each other out so that if something lost electrical charge, another substance would instantly gain the amount being cast away. His work on electricity climaxed in his now famous kite experiment of 1752. Believing lightning to be a form of electricity, and in order to prove it, Franklin launched a kite into a thunderstorm on a long piece of conducting string.

Tying the end of the string to a capacitor, he was vindicated when lightning did indeed charge it, proving the existence of its electrical properties. From these results, and realising the potential of a device that could deflect the harmful effects of lightning strikes away from buildings and property, he developed the lightning conductor.

Franklin had also published his text *Experiments and Observations on Electricity, made at Philadelphia in America* in 1751, which went on to inspire future scientists in the study and development of the uses of electricity.

▸ A PROLIFIC INVENTOR

From 1753 the time Franklin dedicated to science reduced dramatically due to his taking up a new post as deputy postmaster general and, later, political and ambassadorial roles. He did, however, leave a legacy of other inventions from the wide range of experiments conducted throughout his life, including: an iron furnace 'Franklin' stove (still in use today), bifocal spectacles, the street lamp, the rocking chair, the harmonica, an odometer and watertight bulkheads for ships. Franklin also came up with the idea of Daylight Saving Time and was the first to charter the Gulf Stream from observations made by sailors. A man of many talents, Benjamin Franklin was a successful inventor, politician, printer, oceanographer, ambassador, journalist, satirist and, of course, scientist.

THE LEGACY OF BENJAMIN FRANKLIN

Franklin's legacy, in addition to the many inventions such as lightning conductors, bifocal lenses and street lamps, was one of learning. He established one of the first public libraries, as well as one of the first universities: Pennsylvania, in America. On a broader societal level, he established the modern postal system, set up police and fire fighting departments and established the Democratic Party.

He certainly lived up to his own quotation, 'If you would not be forgotten as soon as you are dead and rotten, either write things worth reading, or do things worth the writing.'

JOSEPH BLACK

1728–1799

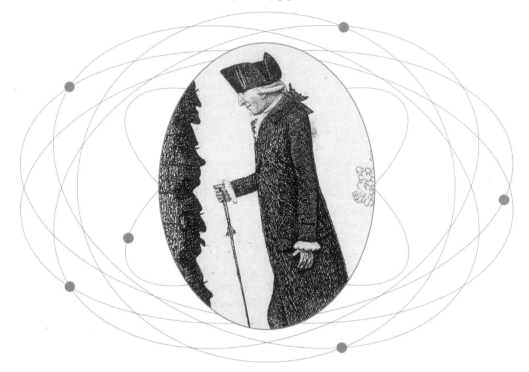

CHRONOLOGY • **1746–50** Black studies chemistry at Edinburgh University • **1754** Presents thesis on the cycle of reactions in chemistry • **1756–66** Professor of Medicine and lecturer in chemistry at Glasgow University • **1757** Discovers the concept of latent heat • **1766** Appointed Professor of Chemistry at Edinburgh, a post he holds until his death in 1799

Joseph Black was born in Bordeaux, France, the son of a wine merchant, and educated at Belfast and later Glasgow universities. At Glasgow he came under the tutelage of the renowned William Cullen, although their relationship soon became that of professor and assistant, rather than that of teacher and student. Cullen was an innovator in the fields of chemistry and the classification of diseases, identifying four major divisions, although he was better known for his unorthodox teaching methods and inspirational lecturing

style. His lectures certainly seem to have inspired the young Black, who was to put chemistry onto a sound, scientific footing, although he never published his works during his lifetime.

▶ THE REDISCOVERY OF CARBON DIOXIDE

Although Jan Baptista **van Helmont** had identified the existence of gases distinct from air more than a century before Joseph Black became prominent, little work had been undertaken to build on these observations in the intervening

Latent heat: the ability of matter to absorb heat while remaining at the same temperature

hundred years. It is for this reason the Scottish scientist is often credited as the discoverer of carbon dioxide, which he called 'fixed gas', even though van Helmont had clearly been aware of its existence. What is true, however, is that Black was the first to fully understand and quantify the properties of carbon dioxide, and in so doing laid one of the key foundations of modern chemistry.

▶ THE IMPORTANCE OF METHOD

Black's insistence on the importance of quantitative experiments was another notable step towards the setting of the standard for the new era of chemistry. He employed such methods in producing the results of his most significant text, *Experiments upon Magnesia Alba, Quicklime, and some other Alcaline Substances* (1756). He outlined the cycle of chemical changes in what has become one of the defining experiments in the teaching of the science. Black observed how limestone (Calcium Carbonate) produced quicklime (Calcium Oxide) and fixed air (Carbon Dioxide) when heated. He then mixed the quicklime with water to produce slaked lime (Calcium Hydroxide), and by combining the output of that with fixed air he was able to make limestone again (plus water). Anticipating the discoveries of Antoine **Lavoisier** a century later, Black concluded from his experiments that carbon dioxide was a distinct gas from 'normal' (atmospheric) air, as well as one of its constituents in small quantities. He also demonstrated that

removing carbon dioxide from the limestone made the latter more alkaline, with the effects reversed on the addition of carbon dioxide again, observing therefore the gas's acidic properties. Black was also able to prove carbon dioxide was made by respiration, the burning of charcoal and through fermentation, but that the gas would not allow a candle to burn in it nor sustain animal life.

▶ THE PHYSICS OF HEAT

Black later turned his attention to physics and here too he made some elementary discoveries now at the core of the subject. Through meticulous experimentation and measurement of results, the scientist discovered the concept called 'latent heat', or the ability of matter to absorb heat without necessarily changing in temperature. The best example of this principle is that of the transformation of ice into water at 0°C. This requires heat to form water, although the liquid formed is still at the same temperature. The same principle is true in the process of transforming water to steam and, indeed, all solids to liquids and liquids to gases. Through this work, Black made the important distinction between heat and temperature. As well as its more general application since, the experiments in latent heat became important almost immediately, as one of Black's friends was James Watt, who benefited from these discoveries during his development of the condensing steam engine.

FURTHER ACHIEVEMENTS

Black also articulated many other findings involving heat. In particular, he formulated the theory of specific heat, that is, the theory which states that different quantities of heat are required to bring equal weights of different materials to the same temperature. Out of this

work came the development of calorimetry, an accurate way of measuring heat for the first time and still used in an amended form today, as well as a device to be employed to this end, the calorimeter.

HENRY CAVENDISH

1731–1810

CHRONOLOGY
• **1731** Cavendish is born in Nice, France, to an aristocratic family
• **1753** Leaves Cambridge University without taking a degree
• **1798** Publishes his estimate of the density of the earth, an estimate almost precisely what it is now believed to be • **1871** The endowment of the famous Cavendish Laboratory was made to Cambridge University, by Cavendish's legatees.

If ever a person were to fit the stereotypical image of a wacky, eccentric scientist, Henry Cavendish would be that man. Born of the English aristocracy and inheritor of a huge sum of money mid-way through his life, Cavendish used his wealth to indulge his unusual behaviour. He built private staircases and entrances to his homes in London so he would not have to interact with his servants, and only communicated with them through written notes. He never spoke to women, doing all he could to avoid having to look at them, and only usually appeared in public for the purposes of attending scientific meetings. His love of solitude did, however, offer him plenty of time to work on the experiments which would advance science, in spite of his equally eccentric approach to the publication of his work.

▶ PROMPTED BY CURIOSITY

Cavendish's main motivation was not scientific acclaim, but curiosity, and it is because of this that he failed to put many of his discoveries into print. He conducted meticulous experiments in

Some of Cavendish's discoveries are considered to be half a century ahead of their time

both physics and chemistry, but it is largely for his work in chemistry that he is best remembered, since he did publish a number of papers in this field.

Of the most famous were his 1766 *Three Papers Containing Experiments on Factitious Airs (gases made from reactions between liquids and solids)*. In these he demonstrated how hydrogen (inflammable air) and carbon dioxide (fixed air) were gases distinct from 'atmospheric air'. Although Joseph **Black** was making similar discoveries with fixed air, Cavendish is credited with being a pioneer in distinguishing and understanding inflammable air. He managed to develop reliable techniques for weighing gases and, in further experiments undertaken around 1781, he discovered that inflammable air, mixed with what we now know as oxygen (from atmospheric air) in quantities of two to one respectively, formed water. In other words, water was not a distinct element but a compound made from two parts hydrogen to one part oxygen (now famously expressed in chemistry as H_2O). Due to his typical tardiness in publication (he did not declare his findings until 1784), his claim to this discovery became confused with similar observations subsequently made by Antoine **Lavoisier** (1743–94) and James **Watt** (1736–1819). The important point is that water was proved not to be a distinct element – a view held since the time of **Aristotle**. In the same paper, Cavendish also explained his discovery that air (whose composition remained constant from wherever it was sampled in the atmosphere) was composed of approximately one part oxygen to four parts nitrogen. In these experiments – performed to decompose air by 'exploding' it with electrical sparks – he also found that there was always a residue of about one per cent of the original mass which could not be broken down further. This 'inert' gas would not be studied again for a century, when it was named argon. In the same series of experiments, Cavendish also discovered nitric acid, by dissolving nitrogen oxide in water.

▸ AHEAD OF HIS TIME

Potentially, Cavendish could have been remembered as a great physicist as well, since some of his experiments and discoveries were considered to be more than half a century ahead of their time. Almost all of his work in this arena remained unpublished until the late nineteenth century however, when his notes were found. The scientist James Clerk **Maxwell** (1831–79) dedicated himself to publishing Cavendish's work, a task he completed in 1879. But by then Cavendish's potential breakthroughs, significant at the time, had been surpassed by history. In particular, Cavendish had undertaken significant work with electricity, anticipating laws later named after their 'discoverers' Charles **Coulomb** (1736–1806) and Georg Ohm (1789–1854), as well as some of Michael **Faraday**'s (1791–1867) later conclusions. In the absence of any other appropriate device and in keeping with his eccentric tendencies, he even resorted to measuring electrical current by grabbing electrodes and estimating the degree of pain it caused him!

THE DENSITY OF THE EARTH

One physical experiment for which Cavendish was acclaimed in his time (and which is now named after him) was working out the density of the earth. The experiments involved a torsion balance and the application of Newton's theories of gravity. In 1798 he concluded that the earth's density was 5.5 times that of water, a figure almost identical to modern estimates.

JOSEPH PRIESTLEY

1733–1804

CHRONOLOGY • **1766** Priestley meets Benjamin Franklin, who awakes his scientific curiosity • **1767** *The History and Present State of Electricity* published • **1771** Discovers that a plant will replenish enough air in a jar to sustain a burning candle • **1774** Discovers oxygen independently of Karl Scheele (who announces his discovery in 1777)

Joseph Priestley was not first and foremost a scientist, yet he became one of the most important British experimental chemists of the eighteenth century. Trained as a Unitarian minister and with an active interest in politics, philosophy, history and languages, it was not until he met Benjamin Franklin in 1766 that his scientific fascination was awakened. The consequent journey this would take him on made Priestley famous, even though he remained an amateur in science for the rest of his life and dedicated most of his time to teaching and preaching.

▸ **BEGINNINGS IN ELECTRICITY**

Physics, not chemistry, was the subject of Priestley's first scientific endeavour, however. Encouraged by Franklin, and given the use of his books, Priestley wrote *The History and Present State of Electricity* (1767) which, as well as acting as a summary of everything known about electricity to that point, included some of Priestly's own contributions such as the discovery that graphite conducts electricity.

▸ **THE ALLURE OF CHEMISTRY**

Physics did not sustain his long-term scientific

Priestley's scientific work virtually ceased after his enforced emigration to the United States

interest, however, and Priestley became more intrigued with chemical experiments. On taking up a new ministerial post in Leeds in 1767, he gained access to a virtually unlimited supply of 'fixed air' (carbon dioxide) with which to begin his work. This he obtained from the gas released through fermentation at his local brewery. Among other things, this led to Priestley's creation of soda water (carbon dioxide in water). Little did he know the consequences this would have for the future development of the soft drinks industry! The episode stimulated the Englishman's interest in working with gases, leaving him determined to add to the mere three that were then known: 'fixed air' (carbon dioxide), 'inflammable' air (hydrogen) and atmospheric air. By improving the design of a piece of apparatus called the 'pneumatic trough', filling it with mercury and then heating solids floating in it, he was able to isolate and capture gases above the mercury. Soon Priestley had discovered four new gases that we now know as nitrous oxide (laughing gas), nitrogen dioxide, nitric oxide and hydrogen chloride.

▶ DISCOVERING GASES

Arguably Priestley's most successful period, however, came while under the patronage of Lord Shelburne at his estate in Calne, Wiltshire, from 1773–80. Ostensibly employed as a librarian and teacher to Shelburne's children, Priestley was largely given the freedom to pursue his scientific studies as he wished. In due course he discovered nitrogen, carbon monoxide, sulphur dioxide, ammonia and silicon tetrafluoride. But it was his discovery of the gas now known as oxygen for which Priestley is most famous.

▶ THE DISCOVERY OF OXYGEN

Priestley stumbled across oxygen in 1774 while heating mercury oxide and seeing that it greatly enhanced the burning of a candle's flame, prolonged the life of mice and was 'five or six times as good as common air.' (Karl **Scheele** had also discovered oxygen independently in 1772 but did not publish his results until 1777). Priestley's findings were in addition to his 1771 discovery that a plant will replenish the 'air' in a jar sufficient to burn a candle again after the candle has burnt itself out. As well as his later observation of the significance of sunlight in the growth of plants, this was an important foundation for other scientists in subsequent research into photosynthesis. Yet despite this, Priestley did not realise the true impact of his discovery and it was left to Antoine **Lavoisier** (1743–1794), whom he told of his findings in 1775, to establish the central place oxygen has in the fields of chemistry and biology. Instead, Priestley named his gas 'dephlogisticated' air, in keeping with the accepted theory that all flammable substances contained the elusive substance 'phlogiston' which was central to the combustion process and released during it.

AFTERWORD

Priestley also undertook other experimental work involving the density, diffusion and heat conductivity of gases, as well as the impact of electrical discharges upon them.

His scientific work virtually ceased, however, following his emigration to Pennsylvania, USA

in 1794. This was a move Priestley felt forced to make after his Birmingham laboratory was subjected to mob violence as a result of his vocal political support for the French Revolution, which he saw as an antidote to the corruption of an un-Godlike society.

JAMES WATT

1736–1819

CHRONOLOGY • **1764** Watt finds a model of Newcomen's steam engine to be inefficient • **1765** Has the idea that kick-starts the Industrial Revolution •
1768 He produces the first prototype of his new engine •**1788** Invents the 'centrifugal governor',
a mechanism that automates speed control • **1790** Perfects the 'Watt Engine'

James Watt is often mistakenly perceived by many people to have been the inventor of the first steam engine. In reality Thomas **Newcomen** had achieved this nearly a quarter of a century before Watt was even born. Watt's engines, however, had the wider impact. Newcomen's machines had been restricted to the world of mining, Watt's were used across all industries. If Newcomen is remembered as the inventor of a power source which changed the world, it is Watt who made its potential available, and provided the catalyst for the Industrial Revolution in the process.

▸ A HAPPY ACCIDENT

As with all the best tales of discovery and invention, the occurrence which began the chain of events leading to Watt's engine was nothing more than a happy accident. In 1764, Watt was asked to repair a scale model of Newcomen's engine which had been used by the University of Glasgow for teaching purposes. The close examination of the model Watt undertook in the process of fixing it made him realise it was hugely inefficient. The biggest weakness Watt identified was in the heating and cooling of the engine's cylinder during every stroke. This

Watt's development of the rotary engine brought mechanisation to industry

wasted unnecessary amounts of fuel, as well as time, in bringing the cylinder back up to steam producing temperature which limited the frequency of strokes.

Consequently, Watt began pondering on improvements to the design of Newcomen's engine. It is said that the Scotsman hit upon his solution in 1765 while wandering through Glasgow Green, and today a memorial stone marks this spot as the place where the idea was born which truly sparked the Industrial Revolution. He had realised the key to improved efficiency lay in condensing the steam in a separate container, thereby allowing the cylinder and the piston to remain always hot.

▶ WATT'S PARTNERS

By 1768, Watt had constructed a fully functioning prototype of his new engine, at which point he entered into a business partnership with John Roebuck to finance and sell the production of the machine. Shortly afterwards, the partnership took out a patent for the engine under the title 'A New Invented Method of Lessening the Consumption of Steam and Fuel in Fire Engines,' and began selling it to colliery owners. Unfortunately, in 1772, Roebuck went bankrupt, although this later gave Watt the opportunity to enter into a more fruitful partnership with businessman Matthew Boulton in 1775.

'Boulton & Watt' immediately applied to the British Parliament for a new patent allowing the company to be the sole makers and sellers of Watt's engines in the country for the next twenty-five years. The success of the application gave the business a virtual monopoly in steam engine production, guaranteeing its financial success and the individual wealth of Watt himself by the time he retired in 1800.

The patent did not stop the inventor from trying to make continuous improvements to his engine, however, and it was not until 1790 that he had finally perfected the 'Watt Engine'. In the intervening years, he made the breakthrough of modifying the steam engine to work in a rotary-motion. The up-down action of his and Newcomen's engines before then had been fine for pumping water out of mines, but of little use elsewhere. With a circular, rotary-motion however, other industries could make use of steam power for driving machines. For example, in the cotton industry, Richard Arkwright was the first to realise the engine could be used to spin cotton, and later in weaving. Flour and paper mills were other early adopters and in 1788 steam power was used to paddle marine transportation for the first time. In that same year, Watt developed the 'centrifugal governor' to regulate the speed of the engine to keep it constant, itself an important foundation in the science of automation.

WATT'S INFLUENCE

Watt is also credited with a number of other inventions including the rev. counter and early letter copying press. More significantly, he was the first to coin the term 'horsepower' which he used when comparing how many horses it would require to provide the same pull as one of his machines. In 1882 the British Association also named the 'watt' unit of power in his honour, further cementing the inventor's fame.

Watt's steam engine was the driving force behind the Industrial Revolution and his development of the rotary engine in 1781 brought mechanisation to several industries such as weaving, spinning and transportation.

CHARLES DE COULOMB
1736–1806

CHRONOLOGY • **1777** De Coulomb publishes a paper outlining the principles behind the construction of an extremely sensitive torsion balance • **1781** Elected to the French Académie des Sciences • **1785** Publishes the principle that becomes known as Coulomb's Law • **1802** Appointed as an inspector of public instruction

Charles Augustin de Coulomb came from a family eminent in the law, in the Languedoc region of France. After being brought up in Angoulême, the capital of Angoumois in southwestern France, Coulomb's family moved to Paris., where he entered the Collège Mazarin. Here he studied language, literature, and philosophy, and he received the best available teaching in mathematics, astronomy, chemistry and botany, before going on to study engineering . The study of electricity was gaining important new ground throughout the eighteenth century, but scientists were still only just beginning to understand how it behaved and could be manipulated, and more importantly, how it could be used. **Newton's** law of gravitation had been a remarkable breakthrough in comprehending the way in which the universe worked, so imagine the impact of the discovery that an identical principle could be applied to electrical forces. Coulomb is credited with this finding.

▸ A FINE BALANCE

Before Coulomb could prove such a phenomenon, however, he would be forced to invent an

Coulomb believed electricity and magnetism to be two distinctly separated 'fluids'

extremely sensitive torsion balance for taking measurements in his electrical experiments. He outlined the principles behind this instrument in a paper of 1777, and went on to construct a balance which could sense force down to as little as 1/100000th of a gram. The appliance itself consisted of a large, glass cylinder with degrees marked around its circular edge, and a wax covered straw inside it, suspended parallel to the ground on a piece of silk thread. An electrically charged ball could be fixed to one end of the straw, balanced by a counter-weight at the other end. By holding a second charged ball at various distances from the one in the cylinder, Coulomb could measure the impact of the electrical force by the degree in which the sphere suspended on the straw rotated.

▸ COULOMB'S LAW

Henry **Cavendish** (1731–1810) had also hinted at the law, but had never published. In its simplest form, Coulomb's Law states that the force between two electrically charged bodies is linked to the square of the distance between them in an inversely proportional relationship. So, for example, by tripling the distance between the charges, the force would decline by nine times. Equally, the force is directly proportional to the product of the charges. In other words, Newton's law of gravitation was mirrored in electricity. Coulomb's law was published in 1785 in one of a series of seven papers the Frenchmen wrote between that year and 1791 outlining the results of his observations. In these ongoing experiments Coulomb also found a similar principle linking the relationship of magnetic forces, leading to speculation by others that perhaps gravity, magnetism and electricity had some kind of interconnected relationship. Coulomb himself, however, had no time for such conjecture, believing electricity, and magnetism in particular, to be two separate 'fluids'. It was left to Hans Christian Oersted (1777–1851), André-Marie **Ampère** (1775–1836) and most notably Michael **Faraday** (1791–1867) to enunciate the phenomenon of electromagnetism.

▸ A MILITARY ENGINEER

While Coulomb is mostly remembered for his electrical studies, he also made discoveries in other areas. The Frenchman spent much of his life as an engineer in the army, working on various French West Indian possessions, as well as a significant amount of this time designing and overseeing the building of fortresses. Hardly surprising, then, that much of his early scientific speculation was linked to engineering theory, such as the concept of a 'thrust line', still used today in construction. The SI unit of electric charge, one unit of which is shifted when a current of one ampere flows for one second, is named after him: the coulomb.

FRICTION BURNS

Coulomb is credited by many commentators with the invention of the science of friction. During his work as a military engineer the issue of friction frequently arose. It was this that inspired him to devote several years of study to the subject. The end result was an articulation of Coulomb's Law of Friction, which outlined a proportional relationship between friction and pressure, and it was this work for which he was elected to the mechanics section of the Académie des Sciences in 1781.

JOSEPH MONTGOLFIER

1740–1810

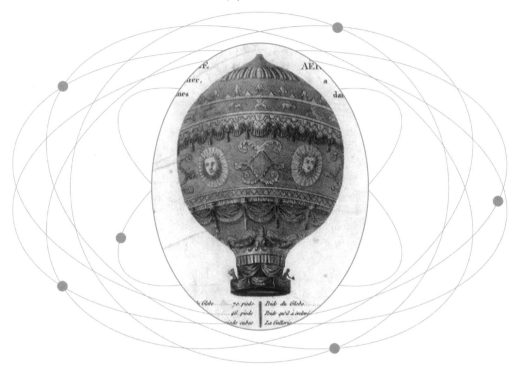

CHRONOLOGY • **1782** The Montgolfiers begin their quest to see if balloons containing heated air can be used to lift humans • **4 June 1783** First public demonstration of their hot air balloon at Annonay • **19 September 1783** Their balloon carries a sheep, duck and rooster to King Louis XVI • **21 November 1783** The first manned flight in history takes place.

The Montgolfier brothers, Joseph and Etienne, were part of a family of sixteen siblings who grew up near Lyons where their father owned a paper-making factory. They had noticed that when paper was burnt on an open fire the hot air would often force the burnt pieces upwards. Around 1782 the brothers began to examine whether this fact could somehow be used in the quest to get humans airborne. They began experimenting, not only with hot air, but also with hydrogen gas (which their countryman

Jacques-Alexandre-César Charles later made use of in producing the first hydrogen balloon) and steam, to see which would raise small paper models the most efficiently. While they struggled to make progress with hydrogen and steam, towards the end of 1782 they succeeded in making paper envelopes filled with hot air rise to the ceiling of their residence.

▸ **FIRST FLIGHT**

From this initial success, the brothers were spurred on to devise a large-scale balloon. By 4

From burning paper, the brothers were spurred on to devise a large scale balloon

June 1783 they were ready. They had produced a cloth sphere with a paper lining some twelve yards across. Inflating the balloon using heat from a fire burning wool and hay, they released it in front of spectators at Annonay, near Lyons. The flight was a success: their invention rose some 2000 yards into the sky in an ascent lasting ten minutes.

▸ A ROYAL AUDIENCE

Word of the brothers' achievement reached Paris where King Louis XVI requested that a display of the balloon should take place at Versailles. The Montgolfiers agreed and upped the ante by not only building a newer, larger balloon, but this time sending it airborne with a sheep, duck and cock as cargo. As well as providing an impressive demonstration, this would also provide proof that living creatures could survive the ascents without adverse effects. The launch took place on 19 September 1783 and although the flight was still short, the balloon travelled an impressive two miles at a height of over 1500 feet. More importantly, the animals touched down unscathed.

The foundations had thus been laid for the achievement of a dream which had burned within mankind for a long time: flight. The volunteers who would make the journey into the unknown were friends of the Montgolfiers, Jean-François Pilâtre de Rozier and François Laurent, the Marquis d'Arlandes. Joseph and Étienne went away again and constructed an even larger balloon capable of carrying human weight. The new balloon was built with a furnace to enable the voyagers to maintain their altitude.

▸ MANNED FLIGHT

The day history was made was 21 November 1783. The human guinea pigs were launched into a twenty-five minute flight across Paris. Even though a large crowd had assembled to watch the launch, the Marquis D'Arlandes later commented, 'I was surprised at the silence and the absence of movement which our departure caused among the spectators, and believed them to be astonished and perhaps awed at the strange spectacle.' Throughout the duration of the flight, the pair rose only a few hundred feet, borne on a hay-fuelled fire, but still gained enough height to avoid the rooftops which sat precariously below. Although the journey was short, they covered several miles and eventually descended into a field on the outskirts of the city, unhurt and triumphant. Man's attainment of the sky saw the beginning of a new era.

Using animals as test pilots was an idea put into practice again when Sputnik II was launched into space by the Russians on 3 November 1957. On board was a dog named Laika, the first canine to enter the cosmos. Unlike the animals that the Montgolfiers used, however, Laika was not destined to survive her journey.

THE MONTGOLFIER LEGACY

For as long as man has observed the birds, he has always dreamt of flying.

From the early Greek fantasies of Icarus's wings, to Leonardo's designs for helicopters and other flying machines, it had been something long talked of, but rarely thought possible. The tale of Icarus was held up as a stern warning to those who sought to better nature. Then came the Frenchman Joseph-Michel Montgolfier and his brother Jacques-Étienne (1745–99). They observed one simple natural phenomenon and sought to take advantage of it. The result was the realisation of the 'unachievable'.

KARL WILHELM SCHEELE

1742–1786

CHRONOLOGY • **1772** Scheele discovers oxygen, two years before Joseph Priestley, but does not publish his findings until 1777 • **1774** Discovers chlorine • **1775** Elected to the Stockholm Royal Academy of Sciences • **1777** *Chemical Observations and Experiments in Air and Fire* published

One of the few challengers to Joseph Priestley's greatness in eighteenth century experimental chemistry had more than just a love of scientific testing in common with him. Like Priestley, the Swede Karl Wilhelm Scheele was only an amateur scientist, had little interest in theoretical science and shared in the claim to the discovery of oxygen. In fact Scheele's claim to be the first is more convincing than Priestley's, given that he made the discovery almost two years before his British contemporary!

▶ DESPITE THE OBSTACLES

Scheele's achievements are all the more remarkable because, unlike Priestley, he did not benefit from a good education, beginning his first full-time apprenticeship at fourteen. Moreover, he was of poor means and throughout his life was forced to conduct most of his experiments restricted by limited apparatus and a lack of laboratory space. His day job was initially as an apothecary and he practised in Gothenburg, Malmo, Stockholm and Uppsala in the years up to 1775. He then moved to Köping to run a pharmacy where he remained for the rest of his life.

Scheele was a man who, quite literally, would die for his science

In the same year as his final move, he was also elected to the Stockholm Royal Academy of Sciences. This recognition brought with it some prestige and consequent well paid job offers overseas, but Scheele preferred to remain in Köping where his limited working hours left plenty of free time to experiment at his leisure without obligation.

▸ OXYGEN BEFORE PRIESTLEY

In 1772 Scheele's scientific efforts resulted in his most significant achievement, the discovery of oxygen, two years before Priestley. He made his 'fire air' from a variety of sources. These including the heating of compounds such as mercury oxide, nitric acid and potassium nitrate. Furthermore, he encompassed his discovery into a more general theory of the atmosphere and combustion, and concluded that the former was composed of only two gases. The first, 'vitiated air,' or nitrogen, suppressed combustion; the other, 'fire air', facilitated it. In common with Priestley, however, he did not appreciate the full significance of his finding, and interpreted it only within the boundaries of the 'phlogiston' theory of his day. But, unlike Priestley, he did not get the credit for it (and even now is often forgotten), because he did not publish his findings until 1777 in his only text *Chemical Observations and Experiments in Air and Fire*. By that point, Priestley was already well known as the discoverer of 'dephlogisticated' air.

▸ A GREEN GAS

The discovery of chlorine was another instance of Scheele not appreciating the significance of what he had achieved. He isolated the green gas in 1774, but it was not until work by others in the early years of the nineteenth century that it was found to be an element in its own right.

▸ A CATALOGUE OF DISCOVERIES

Scheele was obsessed with the discovery of new chemical elements and compounds. From 1770 onwards he identified a remarkable number of substances. These included manganese dioxide, silicon tetrafluoride, barium oxide, copper arsenite, glycerol and hydrogen fluoride, sulphide and cyanide. He also recognised an array of new acids which included citric, arsenic, hydrocyanic lactic, tartaric, prussic and tungstic varieties.

▸ A MARTYR TO HIS CAUSE

Scheele's breakthroughs were not without sacrifice. It is highly likely that the toxicity of many of the chemicals he was working with and his processes for identifying them (including smell and taste tests!) contributed to his relatively early death at the age of forty-three. Scheele was quite literally a man who would die for his science.

AN ARDENT SCIENTIST

Scheele was possibly the most prolific discoverer of new substances that the world has ever seen. This feat is even more remarkable when you consider that he accomplished his achievements in spite of very poor schooling. Taking into account the breadth of his scientific output, his lack of space and adequate laboratory equipment can only reinforce our view of what must have been an iron determination to produce results.

The Swede is also credited with demonstrating the effect that light has on silver salts, a phenomenon that would later became the basis for photography.

ANTOINE LAVOISIER

1743–1794

CHRONOLOGY • **1784** Lavoisier meets the English chemist Joseph Priestley in Paris • **1788** names oxygen • **1789** *Traité élémentaire de chimie* (*Elementary Treatise on Chemistry*) published • **1794** Lavoisier guillotined in Paris

Despite possible claims to the title by other scientists, the Frenchman Antoine Lavoisier is regarded by most as the true founder of modern chemistry. Although he often undertook similar work to that of Henry **Cavendish** (1731–1810), Joseph **Priestley** (1733–1804) and Karl **Scheele** (1742–86), it was the interpretation of his findings which distinguished Lavoisier. His conclusions led to the restructuring of chemistry into a format which laid the foundations for the modern era, arguably achieving an impact comparable to that of **Newton** (1642–1727) in physics. For this it would

be reasonable to assume the scientist could have expected accolades and awards from his countrymen. Instead, they chopped off his head.

▸ **THE CONSERVATION OF MATTER**

Lavoisier's early studies involved experiments concerning the weight loss or gain in substances when heated. By burning matter such as lead and phosphorus in closed vessels, and accurately weighing them before and after heating, he was able to observe the containers did not gain or lose any mass at all during combustion. This ultimately led to his conclusion of the law of conservation of matter. Lavoisier suggested matter

France's most outstanding scientist of his day, Lavoisier met a tragic end on the guillotine

was simply rearranged on heating and nothing was actually added or destroyed overall, hence the equal weight of the vessel before and afterwards. This in itself called into question the 'phlogiston' theory, the commonly held belief all combustible material contained a mysterious element which was released (and lost) on heating.

▶ COMBUSTION THEORY

While the overall weight of the vessel remained the same during Lavoisier's experiments, he made the further interesting discovery that the solids being heated could in fact gain mass. He observed such a reaction, for example, in 1772 when burning phosphorus and sulphur. The only logical conclusion, therefore, was the weight gain of the solid had been caused by some kind of combination with the air trapped in the container. This idea was given further impetus when Lavoisier met Joseph Priestley in Paris in 1774, and the latter explained his discovery of 'dephlogisticated' air. While Priestley failed to realise the impact of this new gas, maintaining a belief in the phlogiston theory, Lavoisier repeated the Englishman's experiments to see if this was the source of the weight gain in some solids during heating. By 1778, he had definitively concluded that not only was Priestley's dephlogisticated air the gas from the atmosphere which was combining with the matter but, moreover, it was actually essential for combustion to take place at all. He renamed it oxygen ('acid producer' in Greek) from the mistaken belief the element was also evident in the make up of all acids. He also noted the existence of the other main component of air, the inert gas nitrogen which he named 'azote'.

▶ MODERN CHEMISTRY

The Frenchmen summarised his new order for chemistry in his 1789 book *Traité élémentaire de chimie* or *Elementary Treatise on Chemistry*, sounding the death bell for the phlogiston theory and beginning modern chemistry. By this point, Lavoisier had also concluded oxygen was equally vital in the process of respiration, performing a similar kind of role in the body as it did during carbon combustion, and was at the basis of all animal life. In addition, the text included a list of known elements to date, identification work he had begun with a number of other French chemists in the mid-1880s. This founded in turn the naming process of chemical compounds which remains to this day. Lavoisier had already outlined one such combination in 1783, proving water was a combination of hydrogen and oxygen, becoming involved in confusion with James Watt (1736–1819) and Cavendish over who had made the discovery first.

Regarded today as the father of modern chemistry, Lavoisier's name is still used in the title of the modern chemical naming system.

AN UNHAPPY END

Despite the importance of Lavoisier's scientific achievements, they were not enough to save his life in the aftermath of the 1789 French Revolution. Lavoisier was a prominent figure in French public life and, most significantly, ran a tax-collecting firm. Those running such companies were considered to be enemies of the revolution. A prominent member of the revolutionary leadership, Marat, who had earlier attempted to forge a career in science and had had his work criticised by Lavoisier, used this as a pretext to try him. It ultimately led to the guillotine and a tragic end to the life of France's most outstanding scientist.

COUNT ALESSANDRO VOLTA

1745–1827

• **1775** Volta invents the electrophorus to produce and store static electricity • **1778** Discovers methane gas • **1780** Volta's friend Luigi Galvani discovers that dead frogs' legs twitch when touched by two different metals • **1800** Creates the Voltaic pile, the first battery, which revolutionises the study of electricity

Whilst the study of electricity had begun to make progress through the work of Benjamin Franklin (1706–90) and others, there was still no reliable way of storing and producing a regular electric current. This had hampered ongoing experimentation in the subject, limiting the scope and usefulness of investigations. One scientist fascinated by electricity and determined to overcome this hurdle was Alessandro Volta.

The Italian aristocrat was born in Como, Lombardy, into a family where most of the male line had entered into the priesthood. Science was clearly Volta's calling, however, and in 1774 he became first teacher and shortly afterwards professor of physics at the Royal School in his hometown. Within a year, he had developed his first major breakthrough in the field of electricity with his invention of the 'electrophorus', used in the production and storage of static electricity.

▸ DANCING FROGS

This device brought Volta recognition within his field and in 1779 he was offered the chair of physics at Pavia University, a post he accepted

To gauge the strength of current from his batteries, Volta used his tongue

and would go on to hold for the next quarter of a century Here he continued his electrical investigations, becoming particularly absorbed in the work of Luigi Galvani (1737–98) during the 1780s. Volta's countryman had made a strange discovery during dissection work. He had found that simply by touching a dead frog's legs with two different metal implements, the muscles in the frog's legs would twitch. Through various other experiments, Galvani wrongly concluded it was the animal tissue which was somehow storing the electricity, releasing the substance when touched by the metals.

▸ THE VOLTAIC PILE

Volta, however, was not convinced the animal muscle was the important factor. He set about recreating Galvani's experiments and concluded, controversially at the time, the different metals were the important factor in the production of the current. Indeed, Volta and Galvani had been friends before the former began criticising the deductions his peer had made concerning the importance of the animal tissue. To make matters worse, it was Galvani himself who had sent Volta his papers on the subject for Volta's review and, he hoped, support. Instead, a bitter dispute broke out concerning whose analysis was correct. Although Galvani would not live long enough to see Volta's ultimate rebuttal of his work, the argument was already swinging in the latter's favour by the time of Galvani's death, and he ended his days a disillusioned man.

▸ DRY AND WET BATTERIES

To back up his theory, Volta had begun putting together different combinations of metals to see if they produced any current, even going as far as to use his tongue as an indicator of current strength from the shock they produced! It was, in fact, an important test because he deduced the saliva from his tongue was a factor in aiding the flow of the electric current. Consequently, Volta set about producing a 'wet' battery of fluid and metals. His decisive solution came in 1800, with the 'Voltaic pile', a stack of alternating silver and zinc disks interspersed with brine-soaked cardboard layers. By attaching a copper wire to the ends of this device and closing the circuit, Volta found it produced a regular, flowing electric current. He had created the first battery.

▸ IMPRESSING NAPOLEON

The invention radically improved the study of electricity, facilitating further breakthroughs in the subject by other scientists such as William Nicholson and Humphrey Davy (1778–1829), who made discoveries using electrolysis, and later aided the work of Michael Faraday (1791–1867). Napoleon, who at that time controlled the territory in which Volta lived, invited the scientist to demonstrate his invention in Paris in 1801. He was so impressed that he made Volta a count, and later a senator, of Lombardy, and awarded him the Legion of Honour medal.

FURTHER ACHIEVEMENTS

The volt, the SI unit of electric potential, is named after the Italian. A volt is defined as the difference of potential between two points on a conductor carrying one ampere current when the power dissipated between the points is one watt.

Volta was also the first to isolate methane gas, an achievement made in 1778.

93

EDWARD JENNER

1749–1823

CHRONOLOGY • **14 May 1796** Jenner diagnoses a milkmaid with cowpox. He extracts matter from her pustules to infect an eight-year-old boy • **1 July 1796** Attempts to infect the boy with smallpox. He doesn't contract the disease and becomes the first person to be intentionally vaccinated against it • **1798** *An Inquiry into the Causes and Effects of the Variolae Vaccinae* published

The rapid development of science in the eighteenth century not only changed people's understanding of the world but often affected their ordinary lives. Few scientists would impact all levels of society in the manner of Edward Jenner, who developed the first ever vaccine.

Jenner was a skilled physician, having completed his training as a surgeon in London between 1770 and 1772 under the guidance of the noted surgeon, John Hunter, before returning to his home village of Berkeley in Gloucestershire,

England to begin work as a medical practitioner. As well as becoming a successful doctor, Jenner was a keen observer of nature, in particular of bird migration habits and the behaviour of the cuckoo. He was also interested in medical experimentation, undertaking work on chemical treatments for certain diseases and investigating the causes of angina by human dissection.

▸ **THE SMALLPOX PREDICAMENT**

His big breakthrough came, however, with his experimental work on smallpox. The disease was

Smallpox was a ruthless, endemic disease, killing one in five of its victims

a ruthless, endemic killer in Jenner's time, ending the lives of one in five of those it infected. Survivors, meanwhile, were often left blind or badly disfigured. There was no known cure for the disease and attempts at prevention were limited to a crude form of inoculation known as variolation. This involved an otherwise healthy person deliberately infecting themselves with smallpox from someone who had a mild form of the disease by taking matter from the patient's sores and placing it into open wounds. As small-pox couldn't be caught for a second time, people hoped by this method to give themselves a mild version of the disease and become immune to it for life. Unfortunately this process was risky, with those inoculating themselves frequently contracting a fatal version of the disease.

▶ FIGHT DISEASE WITH DISEASE

In 1796 Jenner attempted a different approach to prevention. He had heard through anecdotal accounts and observed through local outbreaks that milkmaids who had earlier contracted cowpox from their cattle rarely, if ever, seemed to contract smallpox afterwards. Cowpox was a relatively mild disease caught from cow's udders, far less dangerous than smallpox. In May of 1796, one such milkmaid, Sarah Nelmes, came to Jenner's surgery with cowpox. He extracted some matter from his patient's pustules and persuaded a local farmer to allow him to infect his eight-year-old son, James Phipps, with cowpox in an experiment aimed at preventing him from ever contracting smallpox. The boy, as expected, contracted a mild form of cowpox and quickly recovered. Shortly afterwards Jenner attempted to infect Phipps with a lethal dose of smallpox. No disease was contracted. The doctor repeated the attempt on the boy a few months afterwards, and later on other subjects, but smallpox did not develop. The 'vaccination', as Jenner called it (from the Latin for cowpox, *vaccinae*), had been a success. Jenner did not understand the scientific reasons for this but the practical implications were clear.

▶ THE SUCCESS

Jenner communicated his results to the Royal Society in 1797 but they were declined for publication so in 1798 he printed a paper privately: *An Inquiry into the Causes and Effects of the Variolae Vaccinae*. Although the work initially provoked controversy and vacci-nation was not trusted by everyone its success soon became self evident. From the early 1900s it was acknowledged as the best method for preventing smallpox. The British government agreed, awarding Jenner substantial grants for his work. In 1840 the previous preventative for smallpox, variolation, was outlawed and compulsory vaccination of infants was intro-duced in 1853. The impact was dramatic, with deaths by smallpox reduced from 40 per 10,000 in 1800 to 1 in 10,000 by 1900. In 1980, with no reported cases at all, the disease was offi-cially declared extinct.

FURTHER ACHIEVEMENTS

Jenner never sought to enrich himself on the back of his discovery and during the time he spent promoting its use he suffered financially as his private practice was neglected.

He adhered the advice given to him by his mentor John Hunter, 'why think? – why not try the experiment?'.

Jenner was the first to observe that the newly hatched cuckoo, not the adult, is responsible for removing the other eggs from the nest.

JOHN DALTON

1766–1844

CHRONOLOGY • **1793** *Meteorological Observations and Essays* published • **1801** Dalton states his Law of Partial Pressure • **1803** Outlines his atomic theory in a lecture • **1808** *A New System of Chemical Philosophy* published

For much of his life, the primary interest of the English Quaker, John Dalton, was the weather. Living in the notoriously wet county of Cumbria, he maintained a daily diary of meteorological occurrences from 1787 until his death, recording in total some 200,000 entries. Yet, it was his development of atomic theory for which he is most remembered.

▶ DIFFERENT ATOMS

It was around the turn of the nineteenth century that Dalton started to formulate his theory. He had been undertaking experiments with gases, in

particular on how soluble they were in water. A teacher by trade, who only practised science in his spare time, he had expected different gases would dissolve in water in the same way, but this was not the case. In trying to explain why, he speculated that perhaps the gases were composed of distinctly different 'atoms', or indivisible particles, which each had different masses. Of course, the idea of an atomic explanation of matter was not new, going way back to **Democritus** of Abdera (c.460–370 BC) in ancient Greece, but now Dalton had the discoveries of recent science to reinforce his theory. On further examination of his thesis, he realised that not only would it

Dalton's atomic theory was to transform the basics of chemistry and physics

explain the different solubility of gases in water, but would also account for the 'conservation of mass' observed during chemical reactions as well as the combinations into which elements apparently entered when forming compounds (because the atoms were simply 'rearranging' themselves and not being created or destroyed).

▶ ATOMIC THEORY

Dalton publicly outlined his support for this atomic theory in a lecture in 1803, although its complete explanation had to wait until his book of 1808 entitled *A New System of Chemical Philosophy*. Here, he summarised his beliefs based on key principles such as: atoms of the same element are identical; distinct elements have distinct atoms; atoms are neither created nor destroyed; everything is made up of atoms; a chemical change is simply the reshuffling of atoms; and compounds are made up of atoms from the relevant elements. In the same book he published a table of known atoms and their weights, although some of these were slightly wrong due to the crudeness of Dalton's equipment, based on hydrogen having a mass of one. It was a basic framework for subsequent atomic tables, which are today based on carbon (having a mass of 12), rather than hydrogen. Dalton also wrongly assumed elements would combine in one-to-one ratios (for example, water being HO not H_2O) as a base principle, only converting into

'multiple proportions' (for example, from carbon monoxide, CO, to carbon dioxide, CO_2) under certain conditions. Although the debate over the validity of Dalton's thesis would continue for decades, the foundations for the study of modern atomic theory had been laid and with ongoing refinement were gradually accepted.

Prior to atomic theory, Dalton had also made a number of other important discoveries and observations in the course of his work. These included his 'law of partial pressures' of 1801, which stated that a blend of gases exerts pressure which is equivalent to the total of all the pressures each gas would wield if they were alone in the same volume as the entire mixture.

Dalton also explained that air was a blend of independent gases, not a compound. He was the first to publish the law later credited to and named after Jacques-Alexandre-César Charles (1746–1823). Although the Frenchman had been the first to articulate the law concerning the equal expansion of all gases when raised in equal increments of temperature, Dalton had discovered it independently and had been the first to print.

Dalton also discovered the 'dew point' and that the behaviour of water vapour is consistent with that of other gases, and hypothesised on the causes of the aurora borealis, the mysterious Northern Lights. His further meteorological observations included confirmation of the cause of rain being due to a fall in temperature not pressure.

FURTHER ACHIEVEMENTS

John Dalton began teaching at his local school at the age of 12. Two years later he and his elder brother purchased a school where they taught roughly 60 children.

His paper on colour blindness, which both he and his brother suffered from, and which was *known as daltonism for a long while, was the first to be published on the condition. Dalton is also largely responsible for transforming meteorology from being an imprecise art based on folklore to a real science; how much more precise it is nowadays is perhaps debatable!*

ANDRÉ-MARIE AMPÈRE

1775–1836

The Dane, Hans Christian Oersted (1777–1851), was the first to show that an electric current could deflect a magnetic compass needle, thereby proving the long sought-after link between electricity and magnetism. The Englishman Michael **Faraday** (1791–1867) was the first to make real practical use of Oersted's discovery while it was left to the Frenchman André-Marie Ampère to explain the theory behind it. In so doing, he founded the science of electromagnetism, a branch of science which would have profound importance in helping to shape the modern world.

▸ MATHEMATICS

Ampère was first and foremost a brilliant mathematician and it was his skill in this subject which would facilitate his work in electromagnetism. He excelled in mathematics from an early age and he took up his first teaching post in the subject in 1799 in Lyons. In 1802, he became a Professor of Physics and Chemistry at Bourg-en-Brasse reverting back to mathematics as a

The founder of the science of electromagnetism, Ampère did not have the happiest of lives

professor at the Parisian École Polytechnique in 1809. Napoleon had also made him Inspector General of the university system in 1808. While his professional career advanced fairly effort-lessly, Ampère did not have the same fortune in his private life. He had lost his father to the guil-lotine in 1793 in the aftermath of the French Revolution, and his beloved first wife died shortly after their son was born in 1803. He later remarried, but without happiness.

► ELECTRODYNAMICS

It was against this backdrop that Ampère later undertook his ground-breaking work in electro-magnetism, after seeing a demonstration of Oersted's discovery in 1820. Ampère was by no means a prolific scientist, dipping in and out of the subject at will, but once his interest was stim-ulated in an area he could work extremely quickly. Within just seven days of learning about the link between electricity and magnetism, he had already conducted experiments and began making his own presentations on the phenome-non. Over the following months, he started to formulate mathematical explanations behind the relationship. It only later became referred to as electromagnetism; Ampère at this time had chris-tened it electrodynamics.

► AMPERE'S LAW

In particular, Ampère became interested in the impact that one electric current could have upon another. He had noted two magnets could affect each other and wondered, given the similarities between electricity and magnetism, what effect two currents would have upon each other. Beginning with electricity run in two parallel wires, he observed that if the currents ran in the same direction, the wires were attracted to each other, and if they ran in the opposite direction, they were repelled. He then experimented with other shapes of wires to judge the impact, inter-preting all the results mathematically in a bid to find an encompassing explanation for electro-magnetism. This ultimately resulted in Ampère's law of 1827, another addition to the succession of 'inverse-square' laws begun with Newton's law of universal gravitation. This showed that the magnetic force between two electricity-carry-ing wires was related to the product of the currents as well as the inverse-square of their distance apart. This meant that if the distance between the wires were doubled, the magnetic force would reduce by a factor of four.

FURTHER ACHIEVEMENTS

The Solenoid
Ampère also introduced some important devices, inventing the solenoid, a cylindrical coil of wire which becomes an electromagnet when electricity is passed through it. The sole-noid is most commonly used in devices such as bells and valves, where mechanical motion is required.

The Galvanometer
Ampère exploited Oersted's work, devising an early galvanometer which measured electric current flow via the degree of deflection upon its magnetic needle.

The Ampere
In addition, and perhaps most famously of all, the SI unit of electric current, the ampère, is named after him.

AMEDEO AVOGADRO

1776–1856

CHRONOLOGY • **1776** Avogadro born in Turin, northern Italy, of a long line of lawyers • **1800** Takes up the study of mathematics and physics • **1809** Becomes Professor of Physics at the Royal College at Vercelli • **1811** Proposes theory of volume of gas • **1834–50** Appointed Professor at Turin for a second time • **1856** Dies in Turin • **1860** Stanislao Cannizzaro 'rediscovers' Avogadro's theory and is a forceful advocate at a large chemical scientists' conference

Imagine working enthusiastically in scientific research for most of your life and your contribution to the advancement of the subject being a single achievement or theory. Indeed, many scientists invest years of effort to find themselves in this position and are rightly recognised for their one addition. Now imagine, however, your only theory of note being almost completely ignored in your lifetime and for half a century after you originally proposed it, even though it was the one idea scientists had needed for fifty-years to advance their field! Amedeo Avogadro, who found himself in exactly this position, would have had every right to die a frustrated man.

▶ COMBINATION OF ATOMS

The theory the Italian scientist formulated in 1811 involved a way of integrating the apparently irreconcilable hypotheses of Joseph Louis Gay-Lussac (1778–1850) and John **Dalton** (1766–1844). Indeed, the latter had actively

Avogadro's theory, despite its crucial nature, was completely ignored during his lifetime

sought to discredit Gay-Lussac's law of combining volumes. Gay-Lussac had observed that gases always combined in simple, consistent ratios of whole numbers such as 2:1 or 2:3 (and never in fractions), under the same temperature and pressure conditions. Dalton struggled to accept this because he believed, as a base case, that gases would seek to combine in a one atom to one atom ratio (hence believing the formula of water to be HO, not H_2O). Anything else would contradict Dalton's theory on the indivisibility of the atom, which he was not prepared to accept.The reason for the confusion was because at that time the idea of the molecule was not understood. Dalton believed that in nature all elementary gases consisted of individual atoms, which is true, for example, of the inert gases. This is not the case, however, for other gases which naturally exist, in their simplest form, in combinations of atoms called molecules. In the case of hydrogen and oxygen, for example, their molecules are made up of two atoms, described in chemical notation as H_2 and O_2 respectively. Avogadro realised a comprehension of molecules would explain Gay-Lussac's ratios while at the same time not contradicting Dalton's theories on the atom. For example, by this method Gay-Lussac's ratio for water could be explained by two molecules of hydrogen (making four 'atoms') combining with one molecule of oxygen (or two 'atoms') to result in two molecules of water ($2H_2O$). When Dalton had considered water previously, he could not understand how one 'atom' of oxygen could divide itself (thereby undermining his indivisibility of the atom theory) to form two particles of water. The answer, that oxygen existed in molecules of two and therefore the atom did not divide itself at all, was exactly what Avogadro proposed.

▶ AVOGADRO'S LAW

He built on this principle to famously suggest that at the same temperature and pressure, equal volumes of all gases have the same number of molecules. This became known as 'Avogadro's Law.' The principle in turn allowed a very simple calculation for the combining ratios of all gases, merely by measuring their percentages by volume in any compound (which in itself facilitated simple calculation of the relative atomic masses of the elements of which it was made up).

▶ REDISCOVERY OF AVOGADRO

Few scientists (André-Marie **Ampère** (1775–1836) was an exception) accepted Avogadro's speculation, partly due to lack of experimental evidence, until the Italian Stanislao Cannizzaro 'rediscovered' it and vehemently backed its suggestions at a large conference of chemical scientists in 1860. The law was consequently accepted by many present, immediately clearing up the confusion of the previous fifty years concerning atoms and molecules, and the calculation of the relative atomic and molecular masses of elements.

THE INFLUENCE OF AVAGADRO

As we have seen, Avogadro stands out in our collection of great scientists for having produced only a single theory, which was not recognised until after his death. He merits inclusion in this collection due to the contribution his theory made to molecular biology.

Avogadro was finally vindicated in 1860, four years after his death. In his honour, Avogadro's constant, which expresses the number of particles in a single 'mole' of any substance, currently given as $6.022\ 1367(36) \times 10^{23}$, was named after him.

JOSEPH GAY-LUSSAC

1778–1850

CHRONOLOGY • 1816 Serves as joint editor of the *Annales de chimie et de physique* • 1818 Becomes a member of the government gunpowder commission • 1829 Appointed Director of the Paris Mint Assay Department • 1831 Elected to the Chamber of Deputies • 1848 Resigns from his various appointments in Paris and retires to the country

The Frenchman Joseph-Louis Gay-Lussac may not have died for his science but in 1809 he came close. Having prepared large quantities of sodium and potassium following the first successful isolation of the elements by the Englishman Humphry Davy, Gay-Lussac began using them in other chemical experiments, one of which went spectacularly wrong, blew up his laboratory and left him temporarily blinded. Such were the dangers facing the early chemical investigators. But if the risks were high, so were the rewards, as Gay-Lussac's enduring fame testifies.

▸ THE LAWS OF GAS

Although he went on to make many original contributions to chemistry, inheriting Antoine Lauren **Lavoisier's** (1743–94) mantle as the outstanding French scientist of his day, Gay-Lussac's first major contribution was not his own. Instead, in 1802, he brought to the world's attention a chemical law discovered by his countryman Jacques-Alexandre-César Charles (1746–1823) fifteen years earlier, but which his friend had chosen not to publish. Together with **Boyle's** law, the principle now known as Charles' law (although sometimes also named after Gay-

Gay-Lussac's experiments were as noted for their spectacular nature as for their results

Lussac because of his popularisation of it) completed the two 'gas laws'. It stated that a fixed amount of any gas expands equally at the same increments in temperature, as long as it is also at consistent pressure. Likewise, for a decline in temperature, all gases reduce in volume at a common rate, to the point at about –273°C, where they would theoretically converge to zero volume. It is for this reason that the Kelvin temperature scale later fixed its zero degree value at this point. While the law was not Gay-Lussac's own, his experimental proof was more accurate than Charles'. This helped it to gain acceptance when it was finally published (ironically at around the very time John **Dalton** made the same discovery).

One principle which is entirely attributable to Gay-Lussac is his 1808 articulation of the law of combining volumes. Having confirmed by experimental evidence in 1805 that water was made up of one part oxygen and two parts hydrogen (H_2O) and having further proceeded to break down numerous other compounds, Gay-Lussac noted that gases always combined with each other in simple, small numerical ratios (such as 2:1 or 2:3), and never in fractions. At the time the reason for this was not properly understood. Indeed, John Dalton (1766–1844) sought to discredit Gay-Lussac's conclusion because it appeared to conflict with his own theories on the indivisibility of the atom. In 1811 Amedeo

Avogadro (1776–1856) would provide a framework for both men's theories to work in parallel through his distinction of atoms and molecules (although this too was largely ignored until Stanislao Cannizzaro (1826–1910) rediscovered and articulated it in 1860).

▸ A VOYAGE OF CHEMICAL DISCOVERY

Gay-Lussac would spend much of the remainder of his working life engaged in relentless chemical experimentation, either uncovering new compounds and elements, or greatly improving scientific understanding of the properties of other recently discovered substances. Much of this work was carried out in conjunction with his compatriot Louis Thenard. Together they discovered boron and undertook research into the 'new' element iodine, giving the chemical the name by which we now know it. In 1815 they were the first to create the compound cyanogen, discovering that this was the first in a series of related compounds called the cyanides. The duo conclusively proved Lavoisier's assumption to be wrong– that all acids had to contain oxygen.

Later work included detailed investigations into the properties and reactivity of nitrogen and sulphur, as well as research into the process of fermentation. In addition, Gay-Lussac also worked on modernising chemical experimentation techniques and is credited with creating a precise method of volumetric analysis.

FURTHER ACHIEVEMENTS

Gay-Lussac was as famous for the spectacular nature of his investigative work as for the results it produced.

As well as blowing up his laboratory, he had earlier gained fame for dangerous ascents in balloons, conducted in the name of scientific research. He first took to the skies in 1804

with Jean Baptiste Biot (1774–1862), and later on his own (attaining a then world-record height of 23,000 ft/7km) to investigate the air's composition and magnetic force at these altitudes. His results demonstrated that there was no change in either value from measurements taken on the ground.

CHARLES BABBAGE

1791–1871

CHRONOLOGY • **1815** Babbage helps to found the Analytical Society • **1823** Begins work on the machine later known as his Difference Engine No.1 • **1828–39** Lucasian Professor of Mathematics at Cambridge University, a post previously held by Isaac Newton and later Stephen Hawking • **1833** Work on the Difference Engine is abandoned as Babbage runs out of money • **1991** Doron Swade and his team at London's Science Museum complete the construction of a working No.2; it has functioned ever since

On his death in 1871, the *Times* mocked Charles Babbage, his work was virtually unknown to the wider public and those that had heard of it were generally unappreciative. The Reverend Richard Shipshanks, Babbage's harshest critic, wrote in 1854 with unconscious irony, that for all the public money invested in Babbage's work, 'We should at least have had a clever toy'. It had all begun so well for Babbage. Educated at Cambridge, he proved himself to be a brilliant

mathematician, graduating in 1814 and receiving his MA three years later. In 1822, he began designing what was to become the world's first automatic calculator and after a meeting with the Chancellor of the Exchequer in June 1823 he obtained £1500 to fund the creation of his far-sighted vision. It would become known as the Difference Engine No.1 and would dominate the next ten years of Babbage's life.

The mathematician was driven to attempt the building of such a revolutionary concept and

Babbage's machine 'should be used to calculate the time at which it would be of use' Robert Peel

then persisted with it for so long because of his frustration with the alternative: books of mathematical tables written by teams of number-crunchers to help with complicated calculations. Due to human error they were inevitably prone to mistakes. Babbage was a champion of machines and the scientific approach, taking his enthusiasm to the point of eccentricity. He believed that if a mechanical solution to producing complex calculations could be devised, accuracy would always be assured.

▸ THE ANALYTICAL ENGINE

The government's patience finally ran out in 1834. Ironically this happened at the moment, and in part because of, Babbage's greatest conception: the first programmable computer. He called it his Analytical Engine. It was much more than a calculator, rather an all-purpose computing machine similar in concept to the modern computer. His design envisaged 'programs' written using loops of punched cards. It included a reader able to process the instructions they contained, a 'memory' which could store the results, 'sequential control' and other logical features which would become components of twentieth century computers. Babbage approached the government for more money to build his new machine, even though the Difference Engine was still unfinished. He argued that it would be cheaper and more beneficial to build the Analytical Engine than to make the

necessary changes to finish the first machine. With the original still unfinished, the government was reluctant to fund another ambitious project – it had already swallowed up £17,000 of public money. Babbage persisted in making appeals to the government for extra capital. The scepticism reached the highest echelons of government. Robert Peel once commented wryly, when he was Prime Minister, that Babbage's machine should be employed to 'calculate the time at which it would be of use'.

▸ THE DIFFERENCE ENGINE NO.2

In 1842 the government confirmed it was definitely pulling the plug on Babbage's original project (even though no real work had been undertaken on it for nearly a decade) and there would be no money for his new project. Babbage persisted into the 1850s in trying to raise funds for his Analytical Engine but it never got beyond the design stage. By then, the mathematician had also designed the Difference Engine No.2, a much simpler model which used only a fraction of the 25,000 parts the first engine had employed. It was also slightly smaller than the first at six and a half feet tall rather than eight. Again, however, there was no funding from the government and it failed to progress beyond the design stage. To commemorate the bicentenary of his death in 1991, a team from the London Science Museum finally built a working replica of the No.2 based on Babbage's original plans.

FURTHER ACHIEVEMENTS

Babbage founded a number of societies, including the Royal Astronomical Society in 1820. He also made advances in the theory of algebra, and was responsible for or played a part in numerous other inventions, from the speedometer and the locomotive cowcatcher to the standard rail-way gauge and uniform postal rates. In addition, Babbage was the influence behind an Act of Parliament introduced to curtail the rights of street musicians; his Passages from the Life of a Philosopher (1864) includes an interesting diversion on the social ill of street noise.

MICHAEL FARADAY

1791–1867

CHRONOLOGY • 1821 Faraday creates the first electric motor • 1823 Accidentally liquefies chlorine • 1831 Discovers the principle that will lead to the creation of the electric generator, the transformer and the dynamo • 1833 States the basic laws of electrolysis • 1845 Discovers the Faraday effect

Michael Faraday is regarded as one of the greatest experimental scientists of all time. Even Albert **Einstein** (1879–1955) considered him to be one of the most important influences in the history of physical science. Yet the man whose discoveries and inventions, amongst them the electric motor, electric generator and the transformer, were to have such a profound impact on modern life, might not have entered the scientific arena at all but for certain fortuitous events in his youth. The first was his apprenticeship at a bookbinder's when he was thirteen. Here his interest in science and in particular electricity was stimulated upon reading pages from the books he was tasked to bind. The second fortunate incident was his appointment as assistant to the renowned chemist Sir Humphrey Davy (1778–1829), who had remembered the young Faraday attending his lectures. The temporary post soon turned permanent and shortly afterwards Davy took Faraday with him on a grand European tour which gave the young man the rare opportunity to meet and learn from many of the leading physicists and chemists of the day.

Faraday was considered by Einstein to be one of the most influential physical scientists of all

Much of Faraday's early work as a scientist in the 1820s was not in physics, the area which ultimately led to his breakthrough inventions, but in chemistry. In 1823, he became the first person to liquify chlorine, albeit accidentally, while he was conducting another experiment. He quickly deduced how the new form of chlorine had been obtained and applied the process, which made use of pressure and cooling, to other gases. By employing his talent as an outstanding analyst of his own chemical experiments, he also went on to discover benzene in 1825.

▸ THE ELECTRIC MOTOR

Yet it is physical science, in particular his work involving electricity, for which Faraday is best remembered today. As early as 1821, he was able to create the first electric motor after discovering electromagnetic rotation. He had developed Hans Christian Oersted's (1777–1851) 1820 discovery that electric current could deflect a magnetic compass needle. Faraday's experiment proved that a wire carrying an electric current would rotate around a fixed magnet and that conversely, the magnet would revolve around the wire if the experiment were reversed. From this work, Faraday became convinced that electricity could be produced by some kind of magnetic movement alone but it took ten further years before he successfully proved his hypothesis. In 1831, by rotating a copper disk between the poles of a magnet, Faraday was able to produce a steady electric current. This discovery allowed him to go on to produce electrical generators, the transformer (also invented independently at around the same time by an American, Joseph Henry) and even the dynamo: inventions which can truly be claimed to have changed the world!

▸ ELECTRICAL FIELDS

The reason Faraday was able to make such advances was because from early in his career he had rejected the concept of electricity as a 'fluid', an idea that had been accepted up until that time, and instead visualised its 'fields' with lines of force at their edges. He believed that magnetism was also induced by fields of force and that it could interrelate with electricity because the respective fields cut across each other. Proving this to be true by producing an electric current via magnetism, Faraday had discovered electromagnetic induction. He was encouraged by this and went on to explore the idea that all natural forces were somehow 'united'. He then focused on how light and gravity were related to electromagnetism. This led to the discovery of the 'Faraday effect' in 1845 which proved that polarised light could be affected by a magnet. James Clerk **Maxwell** proved that light was indeed a form of electromagnetic radiation, and eventually provided the mathematical expression for Faraday's law of induction.

THE LAWS OF ELECTROLYSIS

Faraday's fascination with electricity and his background in chemistry both found a natural expression in electrolysis, in which he was also to perform ground-breaking work. In 1833 he was the first to state the basic laws of electrolysis, namely that: (1) during electrolysis the amount of substance produced at an electrode is proportional to the quantity of electricity used and (2) the quantities of different substances left on the cathode or anode by the same amount of electricity are proportional to their equivalent weights.

CHARLES DARWIN

1809–1881

CHRONOLOGY • **1831–36** Darwin takes the job of unpaid naturalist aboard HMS Beagle • **1859** Publishes *The Origin of Species* • **1871** Publishes *The Descent of Man* • **1881** Dies and is buried at Westminster Abbey

The spark for Darwin's accomplishments was ignited with the 1831 HMS *Beagle* expedition, which was to chart coastlines in the South Americas and other areas of the Pacific. Darwin, supposedly studying religion at that time, had become increasingly absorbed with natural history and persuaded the Professor of Botany, John Henslow, to put him forward for the post of unpaid naturalist on the Beagle's voyage. He thereby abandoned his university studies. His father, and initially the vessel's Captain FitzRoy, resisted, but he eventually persuaded them to let him take part in the five-year expedition.

▶ THE GALAPAGOS

During the journey, Darwin made many geological and biological observations, but it was his time spent around the Galapagos Islands which would end up having the most significant impact on him. The ten islands are relatively isolated, even from each other, and as such act as a series of distinct observatories through which Darwin could draw comparisons. He noted that the islands shared many species of flora and fauna in common, but that each land mass often displayed distinct variations within the same group of organisms. For example, he famously noted fourteen different types of finch across the

'Man, with all his noble qualities, still bears the indelible stamp of his lowly origin'

islands, notably with different shaped beaks. In each instance the particular beak seemed to best suit the capture of that bird's prevalent food source, whether it be seeds, insects or fish.

Over the ensuing years, and upon his return to England, Darwin pondered on the reasons for the variations in the finches and other plants and animals. He soon surmised that the birds had descended from a single parent species, rather than each springing up independently and thus acknowledged the idea of evolution, a concept which had existed for some time but was not widely accepted. Darwin began looking for an explanation for this evolution. One text which had a particular impact on him was Thomas Malthus's 1798 work *An Essay on the Principle of Population* which Darwin read in 1838. Malthus had been concerned with overpopulation resulting in famine, and the possible competition for food which could ensue. Darwin immediately saw that this could also be applied to the animal world too, where only those best adapted to food collection in their environments would survive. Those that could not compete would die out and the characteristics of the successful animals, which may have occurred in the first place by chance, would be passed on to future generations. As environments changed and animals moved about, success criteria would change, gradually resulting in variations within species, as had happened with the finches. Ultimately new species would also be created.

▶ CHALLENGING THE NOTION OF GOD

Unfortunately, such a hypothesis would challenge the commonly held view of man as the lord of the earth, specifically created and placed upon the planet in God's image, as described in the Bible. Darwin was implicitly suggesting that man had evolved by chance over thousands of years. He correctly anticipated uproar and resistance to his ideas, particularly from religious leaders. Consequently, he kept his theories dark for twenty years while he gathered additional evidence to back up his case.

He finally published in 1858. He did this jointly with Alfred Russell Wallace (1823–1913), whose independent ideas were remarkably similar to Darwin's. They agreed to a joint public declaration of their hypotheses by submission of a paper to the Linnean Society. Darwin followed this up with a more detailed account in 1859 containing evidence he had collected over the previous decades called *On The Origin of Species by Means of Natural Selection*.

The predicted outcry ensued and a fierce debate followed, but Darwin already had a number of friends, particularly Thomas Huxley, known as 'Darwin's Bulldog', who would vigorously defend his ideas. This left Darwin free to follow through further implications of his hypothesis in other works, including the 1871 text *The Descent of Man*, which articulated the idea of the evolution of the human race from other creatures.

THE LEGACY OF DARWIN

Darwin's ideas took a long time to become generally accepted (even today they are not embraced by everyone), challenging as they did all previous conceptions of what it meant to be human. As has been the case with so many scientists, he encountered paticularly fierce opposition from the Church, whose members *preferred the safety of a sacred text to the uncertainties of observation and experiment.*

The idea of evolution through natural selection is, however, at the heart of modern biology. The man who disappointed his father for lack of academic interest had eventually gone on to turn an entire branch of academia on its head.

JAMES JOULE

1818–1920

CHRONOLOGY • 1840 Joule discovers the properties of Joule's Law • 1840s Determines the principle of the conservation of energy • 1849 *On the Mechanical Equivalent of Heat* published • 1852–59 Together with William Thomson (later Lord Kelvin) he describes the Joule-Thomson effect

Well into the nineteenth century, scientists still did not fully understand the proper-ties of heat. The common belief held that it was some form of transient fluid, retained and released by matter, called 'caloric.' Gradually the idea that it was just another form of energy, expressed as the movement of molecules, began to gain ground. It came to be understood in no small part due to the work of an Englishman, James Joule, who contributed much to the founding of the science of thermodynamics in the process.

▶ JOULE THE BREWER

Joule was actually a brewer by trade and not a full time scientist. Working throughout his life at his wealthy father's brewery, he had not bene-fited from a formal education, and indeed never attended university nor held an academic post throughout his life, which makes his findings all the more remarkable.

His interest in the phenomenon of heat, however, led to his father building him a labora-tory near the brewery. The subject would dominate Joule's studies for the rest of his life.

The discoverer of the First Law of Thermodynamics was a brewer by trade

▸ UNDERSTANDING HEAT

Joule began by examining the relationship between electrical current and resistance, and the heat that they produced. In 1840 this led to his first major achievement, the expression of 'Joule's Law' which mathematically determined the link between current and the resistance in the wire through which it passed in terms of the amount of heat given off. This had added important since it effectively meant that one form of energy was transforming itself into another: electrical energy to heat energy for example, and it undermined the concept of the caloric.

▸ HEAT FROM ENERGY OR WORK

Joule pursued this line of enquiry over the following decade. He proved that heat could be produced from many different types of energy or work, including mechanical energy. He proved it in an experiment in which a paddle was turned by a handle in water and the temperature of the water was seen to rise as a result of this work. Indeed, one of Joule's key skills was his ability to quantify the equivalence of different forms of energy. He used his paddle experiment to deduce the amount of mechanical effort needed to be applied to raise the water temperature by one degree Fahrenheit. From this he formulated a value for the work necessary to produce a unit of heat. Afterwards he summarised his results in an 1849 paper *On the Mechanical Equivalent of Heat,* which brought him acclaim.

▸ HEAT IN GASES

Joule went on to study the role of heat and movement in gases. In 1848 he provided the first estimate of the speed at which gas molecules moved. From 1852 until the end of the decade – together with William Thomson, who later became Lord **Kelvin** (1824–1907) – he continued experiments in this area, and described what became known as the 'Joule-Thomson effect'. This demonstrated how most gases actually lose temperature on expansion due to work taking place to pull apart the molecules. It was a discovery put to wide practical use in the growth of the refrigeration industry later in the century.

▸ JOULE'S LAW

In addition to the First Law of Thermodynamics (see below), Joule also discovered the law that bears his name. It describes the conversion of electrical energy into heat, and states that the heat (Q) produced when an electric current (I) flows through a resistance (R) for a time (t) is given by $Q=I^2Rt$.

The SU unit of energy or work, the joule, is also named after him.

THE FIRST LAW OF THERMODYNAMICS

Joule's most significant discovery was a finding he made during the course of his 1840s experiments. He determined what later became known as the first law of thermodynamics: the principle of the conservation of energy. This was a natural extension of his work on the ability of energy to transform from one type to another. Joule contended that the natural world had a fixed amount of energy within it which was never added to or reduced, but which just changed form.

Julius Robert von Mayer (1814–78) and Hermann Ludwig Ferdinand von Helmholtz (1821–94), independent of Joule and each other, came to similar conclusions at around the same time and are also credited with the discovery of the law.

LOUIS PASTEUR

1822–1895

CHRONOLOGY • **1862** *Mémoire sur les corpuscles organisés qui existent dans l'atmosphère* (*Note on Organized Corpuscles that Exist in the Atmosphere*) published. It puts an end to centuries of debate on the theory of spontaneous generation • **1880** An accident by an assistant leads to Pasteur's development of vaccines • **1885** Pasteur successfully uses his rabies vaccine on a nine-year-old boy, Joseph Meister • **1892** Produces a successful vaccination against anthrax

L ouis Pasteur's name is best remembered for his development of the process of 'pasteurisation.' Though Pasteur was a chemist his most significant breakthroughs were in medicine. Indeed, he is considered by many to be the most important figure in nineteenth century medical research. Much of this reputation hinges on his development of a vaccine against rabies. After Edward **Jenner's** (1749–1823) breakthrough of a vaccine against smallpox made at the end of the previous century, little more had been done to take advantage of the potential of this treatment against other diseases. In 1880, however, Pasteur was to recognise and manipulate a chance occurrence that he noticed in his laboratory to finally systemise a scientific approach to the development of vaccines.

▶ BACTERIA IN CHICKENS

Some chicken cholera bacteria had accidentally been left alone for a long period. Pasteur noticed

Pasteur's process contributed greatly to improving the fermentation of wine and beer

that when he injected this into chickens they did not develop, or only suffered a mild form of, the disease normally associated with the bacteria. When he later injected the same chickens with fresh bacteria, they survived, while others which had not received the earlier treatment quickly died. Pasteur drew parallels between this result and the work of Jenner and set about deliberately applying the approach to other diseases.

▶ RABIES AND PASTEURISATION

By 1882 he had successfully produced a vaccination against anthrax, a disease which seldom affected humans, but which could devastate stocks of sheep and cattle. By 1885 he had developed a vaccine, extracted from the spines of infected rabbits, to successfully treat animals for rabies.

Pasteur's apprehension at performing a trial on humans was cast aside when a nine-year old called Joseph Meister was brought to him. The boy had been bitten several times by a rabid dog. Pasteur injected him with the new vaccine and the boy survived. Word of the success spread and the following year over 2500 infected patients received the same treatment. As a result, fatalities dropped to less than 1%. As well as the immediate benefit and fame that Pasteur's development brought him, it also prompted a rush by other scientists to begin searching for new vaccines for other diseases. Several more successes were heralded by the end of the century.

Prior to this, Pasteur had helped limit the spread of tuberculosis and typhoid through the application of his pasteurisation process. This was developed during his studies on the fermentation of milk and alcohol. Through microscopic examination and other experiments he conclusively countered the prevailing argument of the day which held that it was merely a chemical process. Pasteur proved that microorganisms were essential for fermentation to take place. He also found that potentially dangerous microbes existing in milk, such as those which caused tuberculosis and typhoid, could be destroyed by heating the liquid for about thirty minutes at 63°C. This is now known as pasteurisation, still used to treat milk.

During the same period of work, Pasteur also conclusively disproved 'spontaneous generation' theories, which had persisted for centuries. He demonstrated that sterilised fluid not exposed to microbes in the air would remain uncontaminated, whereas the moment the liquid was put into contact with them it became spoiled.

In addition, from 1865 he greatly aided the French silk industry. By analysing diseases which decimated silkworms he eventually provided successful recommendations for their prevention. Pasteur undertook important work early in his career on the discovery of asymmetrical molecules in compounds which did much for the later development of structural chemistry.

Pasteur used a similar process to pasteurisation to improve the success of fermentation in the wine and beer-making industries.

A MEMORY OF PASTEUR

By the time of his death, Pasteur was world-famous and tributes poured in. Perhaps the most dramatic gesture of all, however, came almost a half-century later. The nine-year-old boy, Joseph Meister, whom Pasteur had saved from rabies, went on to become caretaker at the Pasteur Institute (founded in 1888) where the scientist was buried. In 1940 the Nazis arrived in Paris and ordered Meister to open Pasteur's tomb in order to examine it. Meister chose to kill himself rather than comply with the violation.

JOHANN GREGOR MENDEL

1822–1884

CHRONOLOGY • **1856** Begins his pea-plant experiments • **1865** Mendel first articulates his findings • **1866** *Experiments with Plant Hybrids* published • **1868** Elected as abbot of his monastery where his duties prevent him furthering his research

The work of the Austrian monk, Johann Gregor Mendel, would be at the heart of the future development of biology, and founded a new branch of science in its own right. During his lifetime and for some time afterwards, however, his efforts were largely ignored. Only when others started making similar discoveries in heredity and began looking for related studies was it realised that Mendel had got there decades before them.

▶ JUST TWELVE YEARS

The later impact of Mendel's findings, which effectively act as the starting point for the

modern science of genetics, was even more remarkable given the limited amount of his 'career' actually spent researching them. Up to 1856 his time was spent on religious duties or training at his monastery, or in trying to improve his limited early education sufficiently to allow him to pass his teaching examination. Ironically he never succeeded in attaining the qualification, in part due to his lack of success in biology! From 1868 he became abbot of his monastery, located in the modern day Czech Republic, and had to give up the majority of his scientific research. This meant he completed only twelve years of active experimentation.

Mendel, the father of genetics, had his lowest mark in an examination in biology

▶ THE HUMBLE PEA-PLANT

Even the arena for the breakthrough was an unusual one. Mendel's laboratory was the monastery's garden, his subject the humble pea-plant. The monk had been fascinated by what caused the different characteristics of these plants to occur, such as blossom colour, seed colour and height. He decided to undertake a systematic study of when these features occurred in descendent generations. He set about cross-fertilising plants with different characteristics and recording the results.

▶ UNDERSTANDING HEREDITY

Common assumption at the time was that when two alternate features were combined, an averaging of these features would occur. So, for example, a tall plant and a short one would result in a medium height offspring. The statistical results Mendel collaborated, however, proved something entirely different. Across a series of generations of descendents, plants did not average out to a medium, but instead inherited the original features (for example, either tallness or shortness) in a ratio of 3:1, according to the 'dominant' trait (in the example of height, this was the tall characteristic). He explained this by assuming each parent carried two possibilities for any given trait, for example a tall 'gene' (as we now know it) and a short one for height, or a dark gene and light one for seed colour, or gene 'A' and gene 'B' for 'X' trait. Only one gene from each parent would carry into the offspring (now described as Mendel's law of segregation), however, giving four possible combinations: AA, AB, BA and BB. The 3:1 ratio would be achieved because the 'dominant' gene would feature whenever it were present. So if 'A' were the dominant factor, it would occur three times in four, with the 'B' scenario only occurring when a BB result was obtained. He also noted the different pairs of genes making up the characteristics of the pea plant, such as the two causing height, the two causing seed colour, and so on, when crossed occurred in all possible mathematical combinations, independently of each other. This is now described as Mendel's law of independent assortment, and offered him a simple statistical model for predicting the variety of descendents, backed up by ongoing experimental proof.

▶ DELAYED RECOGNITION

Mendel first articulated his results in 1865, and published them in an article of 1866 entitled *Experiments with Plant Hybrids*. He was frustrated that the conclusions were largely ignored in his lifetime and it was only when three other scientists, Hugo de Vries (1848–1935), Karl Erich Correns (1864–1933) and Erich Tschermak von Seysenegg (1871–1962) independently came across similar experimental evidence in 1900 that Mendel's work was rediscovered. Its importance in explaining principles of heredity across all sorts of life forms (although with refinement in some areas) was soon realised, and it was later used to underpin **Darwin's** argument for natural selection too. The science of what is now known as genetics gradually evolved, and Mendel's position as its, albeit unwitting, founder became cemented in history.

MENDEL'S LEGACY

Although he did not gain any recognition for his work on heredity during his lifetime, he was well respected and liked by his fellow monks and townspeople. Nowadays, Mendel is regarded as the father of the study of genetics

JEAN-JOSEPH LENOIR

1822–1900

• **1807** Francois Isaac de Rivaz builds an early internal combustion engine powered by oxygen and hydrogen • **1859** Lenoir demonstrates his electric-spark internal combustion engine • **1860** Patents his engine • **1863** Uses his engine to power a vehicle • **1876** Nikolaus August Otto introduces a superior four-stroke engine

Unbelievably, the origins of the internal combustion engine, today at the heart of the automobile industry, go back to 1680 and the famous Dutch scientist Christiaan **Huygens** (1629–95). In that year he outlined a design for a primitive form of such an engine which used gunpowder to spark and drive its pistons. The design, however, was never constructed and it would be over a century before anyone would come close to reviving the idea. This time it was a Swiss inventor, Francois Isaac de Rivaz, who in 1807 actually built an early internal combustion engine powered by oxygen and hydrogen. He constructed an accompanying vehicle in which to place it, but the design, being largely impractical, was commercially unsuccessful. It is for this reason, then, that the design patented by the Belgian Jean-Joseph Étienne Lenoir in 1860 is regarded as the first viable internal combustion engine, beginning the revolution in the transport industry which would change the world.

▸ THE ELECTRIC-SPARK INTERNAL COMBUSTION ENGINE

Even as child, Lenoir was fascinated by the

'When I am tall, I will make machines, new machines, machines working all alone'

power of mechanical devices. When only twelve years old he is said to have declared, 'When I am tall, I will make machines, new machines, machines working all alone.' Finding little use for his inventive talents in the small town in which he grew up, Lenoir moved to Paris at the age of sixteen. He initially worked with electricity, making a number of breakthroughs in electro-plating as well as electrical devices for use in the railway industry. But it was engines which absorbed Lenoir and, after many years of design and finally construction, he demonstrated his electric-spark internal combustion engine in 1859, patenting it the following year.

Liquid petroleum fuels were not in use at that time, so Lenoir's engine used a combination of uncompressed air and coal gas to power it in a two-stroke design. Although the method was still primitive by today's standards, it was commer-cially successful. Lenoir worked on subsequent improvements and had sold some 500 versions in and around Paris by 1865. By that point he had already demonstrated its versatility, using it to power a boat in 1861 and a land-based three-wheeled vehicle in 1863.

▶ LATER DEVELOPMENTS

Lenoir's designs would not endure for long, though. The Frenchman Alphonse Beau de Rochas patented the now much more common four-stroke internal combustion engine design in 1862. He did not, however, actually build the engine, which is why the German Nikolaus

August Otto (1832–91) became much more renowned for his commercially successful four-stroke model in 1876. It worked using the 'Otto Cycle,' his explanation for the series of actions which took place on each of the strokes to power the engine. On the first induction stroke, the fuel is drawn into the engine, it is then compressed on the second stroke, ignited on the third using an electric-spark and the residue of gases emitted on the fourth. This process drives a piston up and down which, with the appropriate connecting rods, can be used to power a multitude of devices, most commonly, of course, land vehicles.

Otto's engine still used a combination of gas and air to fuel it and it would not be until the pioneering work of Gottlieb **Daimler** (1834–1900) and others that the petrol internal combustion engine would be invented and following it the widespread manufacture of motorcycles and cars. Indeed, Daimler had worked for Otto and been instrumental in the development of the engine attributed to the latter. It was only when he later set out on his own, however, that the automobile revolution truly began.

His engine typically produced about two horsepower, which was not really big enough to power a vehicle of any size and with its fuel mixture of coal-gas and air was only around four per cent efficient. They were of sturdy construction, however, with some models still running perfectly after twenty years of continuous operation.

FURTHER ACHIEVEMENTS

In addition to his engine, Lenoir also developed a number of other inventions. They include an electric brake for trains (1855), a motorboat using his engine (1886), and a method of tanning leather with ozone.

LORD KELVIN

1824–1907

CHRONOLOGY • **1834** William Thomson enters Glasgow University at the age of ten • **1852–59** Together with James Joule he describes the Joule-Thomson effect • **1858** Patents a telegraph receiver, called a 'mirror galvanometer', for use with submarine cables • **1892** Thomson enters the House of Lords and becomes Lord Kelvin

William Thomson, a Scotsman, was clearly destined for great things from a very young age; he entered Glasgow University at the age of just ten! After studying scientific subjects there he went on to Cambridge, graduating in 1845, and the following year was offered a post as chair of natural philosophy (physics). He accepted and remained in the position for over fifty years, a half-century during which he would exercise significant international authority in his field.

▶ THERMODYNAMICS

Working over such a long period gave Thomson the opportunity to experiment with a wide range of subjects, mostly within the sphere of physics. His key influence, however, was in two areas: thermodynamics and electromagnetism. The former involved a large amount of collaborative work with James Joule, the other dominant British authority on the subject during the nineteenth century, who is among those credited with the first law of thermodynamics. The wider acceptance of this conservation of energy principle, which states energy cannot be created or

The Kelvin scale, used mainly for scientific purposes, defines absolute zero as -273°C

destroyed, was in part due to Thomson's systematic description of it in a paper of 1852. Moreover, the Scotsman went on to independently enunciate and publicise the second law of thermodynamics which described the one way nature of heat flow. Heat can only flow spontaneously from a hotter to a colder body and never from a colder to a hotter one. The German Rudolf Clausius (1822–88) also arrived at the same conclusion during a similar period to Thomson. Together with Joule, Thomson discovered the 'Joule-Thomson effect', whereby most gases actually lose temperature on expansion (from a small nozzle) due to work taking place to pull apart the molecules. This finding was of decisive importance in the growth of the refrigeration industry later in the century.

Thomson also attempted to estimate the ages of the Sun and Earth from theoretical work on the rate at which a hot sphere would cool, but his estimates were significantly below modern calculations due to no knowledge of the heat produced by the phenomenon of radioactivity.

▸ THE KELVIN SCALE

One of the reasons Thomson's peerage name, Kelvin, is so widely known today is because of the work he carried out in defining an absolute temperature scale. He undertook theoretical work to predict that –273.16° Celsius is the point at which the molecules in a substance reached their least energy; nothing could become colder below this point named 'absolute zero'. from this he proposed the Kelvin temperature scale, which has the same increments as the Celsius scale, but defines 0 K (degrees kelvin) as absolute zero. 0°C is 273.16 K and the boiling point of water is 373.16 K. The Kelvin scale is widely used by the scientific community today.

▸ ELECTROMAGNETISM

In electromagnetism, Thomson studied the work of **Faraday** (1791–1867) and sought to add to or reinforce his findings. In particular, he attempted to offer some of the mathematic backup Faraday lacked when expounding his theories, helping to gain acceptance for the idea of electromagnetic fields. He also contributed thoughts for the basis of the electromagnetic theory of light, although in this, as with his mathematical work, he would only be partially successful.

It would require James Clerk **Maxwell** (1831–79) to later pull together the work of Faraday and Thomson into a definitive, mathematically sound hypothesis. Thomson did, however, highlight the correct voltages for underwater telegraph signal cables, and was then instrumental in the project which laid the first transatlantic cable, completed in 1866. Due to these successes Thomson became rich and in 1892 was elevated to the peerage, taking the name by which he is best remembered today, Lord Kelvin.

FURTHER ACHIEVEMENTS

William Thomson risked his life by personally taking part in the hazardous process of laying the first transatlantic cable that was to make him his fortune.

He introduced the concept of an absolute zero temperature, a temperature below which nothing can fall.

A man with many interests, Thomson also concerned himself with geophysical questions about tides, the shape of the Earth, atmospheric electricity, thermal studies of the ground, the Earth's rotation, and geomagnetism. He held the chair of natural philosophy at the University of Glasgow for 53 years.

JAMES CLERK MAXWELL

1831–1879

The Scottish physicist James Clerk Maxwell's breakthroughs in electro-magnetism came largely in the early 1860s while he was a professor at King's College, London. He examined Faraday's idea concerning the link between electricity and magnetism interpreted in terms of fields of force and began to search for an explanation for this relationship. Maxwell soon saw that it was simple: electricity and magnetism were just alternative expressions of the same phenomena, a point he proved by producing intersecting magnetic and electric waves from a straightfor-ward oscillating electric current. Furthermore, Maxwell worked out that the speed of these waves would be similar to the speed of light (186,000 miles per second) and concluded, as Faraday had hinted, that normal visible light was indeed a form of electromagnetic radiation. He argued that infrared and ultraviolet light were the same and predicted the existence of other types of wave – outside of known ranges at that time – which would be similarly explainable. The discovery of radio waves in 1888 by Heinrich Rudolph **Hertz** (1857–94) would later confirm this.

'The most profound and fruitful [ideas] since the time of Isaac Newton' *Albert Einstein on Maxwell*

▶ ELECTROMAGNETISM

But Maxwell did not stop there. In 1864, he published his Dynamical Theory of the Electric Field which offered a unifying, mathematical explanation for electromagnetism. The text was based around four equations, now known simply as 'Maxwell's equations,' which outlined the relationship between magnetic and electric fields. He later wrote another piece on this association, published in 1873 under the title Treatise on Electricity and Magnetism.

▶ MAXWELL AND BOLTZMANN

While Maxwell's most outstanding achievements were in explaining electromagnetic radiation, he also undertook important work in thermodynamics and would offer important kinetic explanations for the behaviour of gases. This involved building on the idea of the movement of molecules in a gas. The Scotsman proposed that the speed of these particles varied greatly. Again he used his mathematical skills to produce a statistical model which would reinforce the ideas behind this research, now known as the Maxwell-Boltzmann distribution law (the last part of the name coming from the Austrian Ludwig Eduard Boltzmann who independently discovered the same explanation). Amongst other things, the convincing explanation that heat in a gas is the movement of molecules would finally do away with the caloric fluid theory of heat.

▶ COLOUR PHOTOGRAPHY

Maxwell's other accomplishments involved the deduction that all other colours can be created from a mix of the three primaries. In 1861 he applied this discovery practically in photography, producing one of the first ever colour photographs. Earlier in his career Maxwell had studied Saturn's rings and concluded that they were made up of lots of small bodies and could not be either a liquid or whole solid as had previously been speculated. In 1871, he returned to Cambridge and became the first Professor of Physics at the Cavendish Laboratory, which he helped to establish. The laboratory became world-renowned, dominating the progress of physics for many decades and producing countless leading scientists.

It is highly possible that the Scotsman himself could have led many other breakthroughs had it not been for his tragically early death. He contracted and died from cancer aged just 48.

Although regarded as a slow learner by his first tutors, William Hopkins, one of the country's brightest minds, recognised his great ability at university. **Einstein** also described the change in the conception of reality in physics that resulted from Maxwell's work as 'the most profound and the most fruitful that physics has experienced since the time of **Newton**.'

MAXWELL'S SCIENTIFIC LEGACY

He may not be as well known, but James Clerk Maxwell's standing as a scientist is often considered by many to be on a par with Isaac Newton and Albert Einstein.

Like those other great scientists, he offered explanations for physical phenomena which would revolutionise our understanding of them.

He forged a path for scientists to follow by taking the experimental discoveries of Faraday (1791–1867) in the field of electromagnetism and providing a unified mathematical explanation for an achievement which had evaded other minds for so long.

ALFRED NOBEL

1833–1896

CHRONOLOGY • **1864** Nobel's nitroglycerine factory explodes, killing his brother • **1866** Invents dynamite • **1876** Invents blasting gelatin • **1886** Invents ballistite • **1896** The Nobel Foundation is set up to comply with the terms of Nobel's will • **1901** The first Nobel prizes are awarded

A lfred Bernhard Nobel is a unique entrant in this book. His research and inventions were important but perhaps not significant enough in themselves to demand inclusion in this compilation. However, the fact that his name is indirectly associated with not just one, but scores of scientists who changed the world through the prize named after him, certainly does.

▸ A MAN OF THE WORLD

Although he was a Swede, born and brought up in Stockholm, most of Nobel's education took

place in Russia. His family moved to the country in 1842 to join his father, an engineer, who had taken on a supervisory position in St. Petersburg. During his education there Nobel showed a talent for languages, mastering Russian, French, English, German and Swedish, but it was chemistry which truly captured his interest. In 1850 he went on to study the subject in Paris, before spending a number of years in the United States. He then went back to St. Petersburg until 1859 where he stayed.

When Nobel was not allowed to rebuild his factory, he carried out research on a barge

▶ NITROGLYCERIN

Nobel eventually returned to his native Sweden and began putting his chemical knowledge to practical use. He set up a factory to manufacture the relatively unstable liquid explosive, nitroglycerin, to serve the growing market for it in civil engineering. But in 1864, perhaps predictably, disaster struck. There was a massive explosion, destroying Nobel's factory and, tragically, killing five people, one of whom was his own brother Emil. The accident set Nobel on a determined path to develop a more stable explosive but the government would not allow him to rebuild his factory, so he had to resort to carrying out his chemical research on a barge.

▶ STABLE EXPLOSIVES

In 1866 he made his breakthrough. He found that the liquid explosive became safe to handle if it was absorbed into a substance called kieselguhr and packed into small 'sticks.' Nobel called the invention dynamite and successfully gained patents for it in the United Kingdom and the United States. The relatively safe, yet powerful, nature of the explosive became widely popular and was a commercial success. He went on to improve the effectiveness of his invention, developing a more formidable substance called blasting gelatin in 1876, and another compound ten years later called ballistite. His other inventions included a series of detonating devices.

These replaced the need for a live spark to fire his explosives, further improving their safety.

▶ THE NOBEL PRIZES

The success of Nobel's dynamites as well as his interest in oil helped him to obtain vast personal wealth. Ironically, for a man who had spent most of his life developing explosives, Nobel was a pacifist. Although he hoped the devastating potential of his inventions would act as a deterrent to war, he feared how they might be abused in the future. This was one of the reasons he chose to leave much of his fortune to funding the establishment of a series of awards, one of which included an accolade for peace. There was another dedicated to literature, with the remaining three presented for achievements in the sciences. The first Nobel prizes for medicine (or physiology), physics and, of course, chemistry were awarded in 1901 and since then they have become synonymous with excellence in their related fields. They are awarded on an annual basis, according to the terms of Nobel's will, 'to those who, during the preceding year, shall have conferred the greatest benefit on mankind.' By their very definition the prizes are presented to scientists who have changed the world, and encourage others to endeavour to do so, and as a consequence have established Alfred Nobel as a scientist who did the same.

FURTHER ACHIEVEMENTS

The sixth Nobel award, the prize for economics, was added in 1968 by the Bank of Sweden, and was first awarded in 1969.

Alfred Nobel was prolific as an inventor, obtaining over 350 patents in various countries: artificial leather, silk, and the blasting cap were among his many inventions.

His wealth came not only from explosives, but also from holdings in his brothers' oil companies, as well as extensive involvement with the Swedish arms industry, particularly the Bofors Company. Ironically, for someone who made his name with the ivention of dynamite, Nobel was a committed pacifist.

WILHELM GOTTLIEB DAIMLER

1834–1900

CHRONOLOGY • **1885** Daimler patents the first petroleum-injected internal combustion engine • **1885** Invents the motorcycle • **1886** Invents the four-wheeled petrol-driven automobile • **1889** Constructs an improved two-stroke, four-cylinder engine • **1899** Produces the first Mercedes automobile • **1893** Karl Benz produces the first mass-produced automobile • **1926** Merger of Daimler and Benz

The German Gottlieb Daimler spent much of his life working with engines long before he made the breakthrough which would change the way the world travelled. When success did arrive, however, it followed quickly and dramatically. Daimler had become convinced early that steam power was outdated. A temperamental workaholic, he perfected the first petrol internal combustion engine, and produced the first motorcycle and the first four-wheeled petrol driven car, within only a few years of each other. Daimler

was a cosmopolitan man and founded companies in England and France as well as Germany.

▸ EARLY WORK

The foundations for Daimler's work had already been laid by the earlier pioneering efforts of Jean Joseph Étienne **Lenoir** (1822–1900), Alphonse Beau de Rochas (1815–93) and Nikolaus August Otto (1832–91) in the creation of two and four-stroke gas-fuelled internal combustion engines. Daimler himself had been involved in these early developments as technical director at Otto's

Daimler, the man who made the automobile revolution possible, never liked driving

factory from the early 1870s, and had been instrumental in the development of the four-stoke engine and the 'Otto cycle' which drove it.

▶ THE BREAKTHROUGH

In 1882 Daimler left Otto's employment to set up in business with Wilhelm Maybach (1846–1929), an engineer who had worked for him in the old company and who was undertaking pioneering work on the use of liquid petroleum as a possible fuel source in the internal combustion engine. Although the fuel had been known about for thousands of years and had been commercially available for decades, it had been of no use in the developing internal combustion engine industry because the liquid could not be compressed in the same manner as gas. Together, though, through their development of a carburettor, Daimler and Maybach made the breakthrough which would allow them to take advantage of the fuel source.

The carburettor converted the liquid petroleum into a thin spray which could be compressed and sparked in a four-stroke engine, just like the earlier gas models. The company's first patents were recorded in 1883 and by 1885 they had fully developed a modern-style lightweight petroleum- injected engine which would become the basis for the burgeoning automobile industry.

▶ THE FIRST VEHICLES

Daimler took immediate advantage of his engine, using it to power his 1885 'Reitwagen', the world's first motorcycle. In 1886 he invented the first four-wheeled petrol-driven automobile by using his engine to power a stagecoach. In between times, however, he had missed out on the claim to the world's first internal combustion engine motorcar. Although it was driven by a less advanced 0.75 horsepower engine, Karl Benz (1844–1929), a name also still famous within the automobile industry, had designed and constructed a three wheel vehicle with a superior electrical ignition.

▶ MERGER WITH BENZ

In 1889 Daimler constructed an improved two-cylinder, four-stroke engine. The following year Maybach improved this to a four-cylinder, four stroke version. Benz, meanwhile, was making advances in automobile manufacture and was rolled out the first mass-produced car in 1893, the Benz Velo. Both companies became leaders in the car industry, a position they still hold onto today, and combined forces through a merger in 1926. Daimler had long since passed away but Benz was still alive and served on the new company's board for his remaining years.

Ironically, Daimler's death had been hastened by a car journey in bad weather which he had insisted on taking against doctor's orders. What is more, the man who had made the automobile revolution possible had apparently never liked driving!

DAIMLER, MERCEDES AND BENZ

It was the development of the carburettor that began the age of petroleum powered transport Many of the principles he developed are still used in modern vehicles.

The Mercedes name was originally the name of Emil Jellinek's daughter. Jellinek was a businessman who raced Daimler cars under the pseudonym 'Mercedes'. In 1900 there was an agreement to develop a new engine under the name Daimler-Mercedes. As the engine proved to be virtually unbeatable the name stuck and was registered as a trademark.

A Daimler-powered car won the first international car race, from Paris to Rouen, in 1894.

DMITRI MENDELEEV

1834–1907

CHRONOLOGY • 1860 Mendeleev attends a lecture by Stanislao Cannizzaro which has a profound influence on his later work • 1868–70 The Principles of Chemistry published • 1869 *On the Relation of the Properties to the Atomic Weights of Elements* published, containing the first periodic table • 1893 Becomes director of the Bureau of Weights and Measures • 1955 Element 101 discovered and named mendelevium in his honour

Mendeleev was born the last child of a large family in Tobolsk, Siberia, whose blind father could not support it. His mother enabled the family to survive by setting up and running a glass factory. Under such circumstances Mendeleev's early education was limited although he did attend school. He showed enough promise, though, to encourage his mother to leave Siberia in 1848, shortly after his father had died and the glass factory had burnt down, in a quest to

enable him to enter university. Mendeleev was denied entry to a number of academic schools before settling to study science at the Pedagogic Institute in St. Petersburg. Here he excelled and qualified as a teacher in 1855 before being given additional opportunities over the following years to study at universities in Russia and overseas.

▸ HIS GREATEST INLUENCE

In 1860 Mendeleev attended an important chemistry conference at Karlsruhe where the Italian, Stanislao Cannizzaro (1826–1910), passionately

Discoverer of the periodic table, Mendeleev was to miss out on a Nobel Prize by one vote

announced and backed his rediscovery of the distinction between molecules and atoms (originally made in 1811) by **Avogadro** (1776–1856). Understanding of the atomic weights of elements had been confused for half a century without this distinction and Cannizzaro's speech was a profound influence on the development of Mendeleev's later work.

During the 1860s Mendeleev returned to St. Petersburg, finally becoming Chair of Chemistry at the university in 1866. He became aware during this period that chemistry lacked a comprehensive teaching textbook, so set about writing his own. This was finally published in 1869 under the title *The Principles of Chemistry*, setting a new standard. It was while he was researching this book that Mendeleev returned his attention to the atomic weights of elements, and introduced his card game into the equation.

▸ THE STRUCTURE OF ELEMENTS

Mendeleev wanted to list the known chemical elements in a structured way. Other scientists had tried to do the same in the past but were unsuccessful in finding a uniform way in which to list them or, indeed, in even deciding criteria by which to arrange them. So Mendeleev decided to write the properties of each element on a single card and began placing them in different formations according to various principles. He quickly discovered that if he positioned the elements

according to atomic weight in short rows underneath each other, the resultant columns seemed to share common properties. The British chemist, John Alexander Reina Newlands (1837–98), had independently made a similar observation in 1864 but had had his observations ignored.

▸ THE PERIODIC TABLE

Mendeleev took his work a step further. He drew up a 'periodic table' of the elements according to their atomic weights and the common properties he found in the columns. He realised that for this scheme to work it was necessary to leave spaces for elements which he believed were as yet undiscovered. He could, though, predict their likely properties and weights and was vindicated over the coming years when gallium, scandium and germanium were discovered to slot into the gaps he had left.

Mendeleev also believed that some atomic weights, such as that for gold, had been miscalculated and he re-estimated their details to fit his structure. Again, more accurate measurements would later prove Mendeleev's assumptions to be correct. He first published his table in 1869. The text was not widely accepted at first, but eventually it became the standard method of classifying the chemical elements, restructuring the entire subject of chemistry and greatly aiding scientists of all disciplines in understanding the properties and behaviour of the elements.

MENDELEVIUM, ST PETERSBURG & THE TABLE

Mendeleev predicted three yet-to-be-discovered elements including eke-silicon and eke-boron and his table did not include any of the Noble Gases which were still unknown. Mendeleev also investigated the thermal expansion of liquids, and studied the nature and origin of petroleum. In 1890 he resigned his professor-

ship and in 1893 became Director of the Bureau of Weights and Measures in St. Petersburg.

The element with the atomic number 101 was discovered in 1955 and named mendelevium as a tribute to the great Russian scientist who narrowly missed out on the Nobel Prize for Chemistry in 1906. He lost by one vote.

WILHELM CONRAD RÖNTGEN

1845–1923

CHRONOLOGY • **1868** Röntgen receives his doctorate for his thesis *States of Gases* • **1894** Begins experimenting with cathode rays • **8 November 1895** Inadvertently discovers X-rays • **28 December 1895** Reveals his finding to the world • **1901** Becomes the first person to be awarded the Nobel Prize for Physics • **1912** X-rays finally understood as a result of the work of Max Theodor Felix von Laue

Today the uses of X-rays, particularly in hospitals, are well known to the general public, yet little more than a century ago leading physicists were not even aware of their existence. It would take a chance discovery by a German named Wilhelm Conrad Röntgen to change that and begin a process which would not only result in an understanding of X-rays, but lead on to pioneering work in radioactivity.

▶ 'X' RAYS

Röntgen was a successful scientist in his own right long before he stumbled upon X-rays. He was a Professor of Physics at the University of Würzburg in Germany from 1888 and had conducted research in many areas. But he was largely unknown to the wider world until 28 December 1895 when he unfurled the exciting discovery for which he would subsequently be remembered. The story of the rays Röntgen named 'x', because of their mysterious properties on his first finding them, however, had actually

Exposed to Röntgen's X-rays, bones would appear as shadows against a screen

begun a few weeks earlier in Röntgen's laboratory, on 8 November 1895.

He had been undertaking some tests involving little understood cathode rays when he noticed something unusual. He knew that the cathode rays emitted from the device he was using to project them could only travel a few centimetres, yet he suddenly noticed that another item in the darkened room became illuminated during the test. It was a screen covered in a substance called barium platinocyanide and Röntgen realised straight away that the glow could not have been caused by the cathode rays as the object was over a metre away. He thought that perhaps it indicated some unidentified radiation being emitted when the rays hit the glass wall of the projection device. He began excitedly investigating the properties of his accidental discovery.

▸ PICTURES OF BONES

Before announcing his finding to the world he uncovered many of the properties of the rays, including some of the factors which would go on to make them so useful in the future. For example, Röntgen discovered that the rays would pass through many different kinds of matter including metals, wood and, significantly, human limbs. Indeed, bones would appear as shadows against a screen or photographic plate allowing an X-ray image of them to be taken. He also found that the rays travelled in straight lines and were not knocked off course by electric or magnetic fields. But he was unsure exactly what the rays were; they had some characteristics in common with light rays, but did not reflect or refract like light. It was not until 1912 that they were fully understood, when Max Theodor Felix von Laue (1879–1960) showed that they were a form of electromagnetic radiation with a wavelength shorter than visible light.

▸ MEDICAL APPLICATION

The benefits of X-rays in medicine were quickly brought into common use, and as they were better understood they were applied to other areas, such as the study of the structure of molecules and in researching the properties of crystals. Others scientists ran into new phenomena as a by-product of their researches into X-rays, most notably Antoine-Henri **Becquerel** (1852–1908) who began to understand radioactivity as a result of his investigations. By the same token, it took time for the potentially harmful effects of X-ray radiation to be understood, and Röntgen's health was affected by his experiments.

Röntgen did, however, become an early beneficiary of the Nobel Prize. In 1901 he was the first person to receive the award for physics in recognition of his discovery.

RÖNTGEN'S OTHER ACHIEVEMENTS

Wilhelm Röntgen was the first person to take X-ray photographs. His pictures included amongst other things images of his wife's hand.

After his discovery it only took him six weeks to determine many of the properties of X-rays. The development was to be instrumental in the later discovery of radioactivity.

Röntgen also worked and researched in other scientific fields: elasticity, capillarity, the specific heat of gases, conduction of heat in crystals, piezoelectricity, absorption of heat by gases, and polarised light.

Sadly, as a result of his experiments Röntgen and his technician were both affected by radiation poisoning.

THOMAS ALVA EDISON

1847–1931

• **1870** Edison's first commercially successful invention, the universal stock ticker • **1875** Sets up his laboratory at Menlo Park • **1877** Patents the carbon button transmitter, still used in telephones today • **1877** Invents the phonograph • **1879** Invents the first commercially incandescent light

Amongst the plethora of universally renowned inventions that Thomas Edison is credited with is the expression, 'Genius is one percent inspiration and ninety-nine percent perspiration.' As a summary of the work ethic that the man applied to his own life, it could not be more appropriate.

▶ TRIAL AND ERROR

That is not to undermine the highly creative and original mind that was at the root of Edison's inventions. The fact that his nature and approach

to his science were the complete antithesis to almost every academic and inventor of his time, scorning high-minded theoretical and mathematical methods and relying more often than not on trial and error in practical experiments, was a kind of genius in itself. Commentators have attributed Edison's unique mindset to a host of causes, from his lack of formal education, having left school at twelve, to his increasing loss of hearing from the age of fourteen, ultimately to the point of virtual deafness, which allowed him to focus without distraction on his work. His mother reported that from the earliest age he

'Until man duplicates a blade of grass, Nature can laugh at this "scientific" knowledge'

questioned everything he learnt, but others might just put it down to the 'one percent.'

Edison simply refused to accept that 'impossibilities' could not become facts without relentless experimentation and results which convinced him to the contrary. In Edison's own words, 'I find out what the world needs. Then I go ahead and try to invent it.'

With some 1093 patents singularly or jointly held in his name by the time he stopped inventing at 83, nobody could doubt that Edison meant what he said. He was the most prolific inventor the world had ever known, filing a patent once every two weeks of his working life. Given such a statement, the question is not how he changed the world, but which one of his inventions changed the world the most.

▶ HIS GREATEST WORK

Was it, for example, the phonograph, the first ever sound-recording machine, designed and invented in 1877, surprising even Edison himself when it actually worked? Perhaps it was the first commercially incandescent light bulb, successfully produced in 1879 after more than 6,000 attempts at finding the right filament until landing on a solution in carbonised bamboo fibre. Could it have been the creation of the first commercial electric light, heat and power system, centrally generated to provide power directly into homes and businesses – set up by Edison in

Lower Manhattan 1882 – ultimately leading to the creation of the company General Electric? Or was it the development of devices for recording and playing moving pictures, the Kinetograph and the Kinetoscope respectively, available commercially from 1894, leading on to silent movies and the industry which followed? Then there are the other inventions remarkable in themselves: the carbon button transmitter, still used in telephones today, which made Bell's telephone audible enough for practical and commercial exploitation; the dictaphone; the mimeograph; the electronic vote recording machine, his first patented invention; or the universal stock ticker, his first commercially successful invention sold in 1870 for $40,000, enabling him to fund the research which led to his later inventions.

▶ REVOLUTIONARY APPROACH

Edison's revolutionary approach of establishing dedicated research and development centres full of inventors, engineers and scientists, working day and night on testing and building, brought many of his ideas to fruition. This began with his laboratory in Menlo Park, New Jersey, in 1876. Not only did these centres help Edison practically complete his own inventions but they also changed the rest of the business world's approach to research and development.

THE LEGACY OF EDISON

Arguably the best-known American of his generation, Thomas Edison was considered to be retarded at school due to his hearing difficulty, and attended only occasionally for five years.

Despite this inauspicious start Edison unquestionably changed the world, holding as he did 1093 patents singularly or jointly. Yet

this most prolific inventor felt he had only scratched the surface of the possible. 'Until man duplicates a blade of grass,' he once said, 'nature can laugh at this so called "scientific" knowledge,' adding, 'We don't know one millionth of one percent of anything.'

ALEXANDER GRAHAM BELL

1847–1922

CHRONOLOGY • **1870** Following the death of Bell's two brothers from tuberculosis, the Bell family emigrates to Canada • **1873** Bell becomes Professor of Vocal Physiology at Boston University • **1875** His multiple telegraph is patented • **1876** Bell patents the telephone

Although Alexander Graham Bell had put into practice the previously fanciful notion of voice communication using a wire before he was even thirty, his path towards the invention of the telephone was, physically at least, a long one. A Scotsman, born and brought up in Edinburgh, Bell was mostly taught at home with some limited education at the University of Edinburgh and University College, London. His development of the telephone, however, took place in Boston, in the United States in 1876. Between times, he had

taken up a teaching post in Elgin, Scotland. It was here that he began to study the sound waves which would prove so important in the creation of his revolutionary device.

▶ EMIGRATION

After this worked with his father in London – like his son, Bell senior was a trained speech therapist. Tragedy struck shortly afterwards with the successive deaths of his two brothers from tuberculosis. This acted as the spur for the remaining members of the Bell family to emigrate to Canada in 1870. But by this time Alexander had

The inventor of the telephone, Bell devoted much of his life to working with the deaf

also contracted the disease. He successfully convalesced on arrival in Canada and in 1871 was well enough to move to the US city of Boston which provided the setting, cast and finance for his master creation. He began first, however, by giving a series of lectures on his father's Visible Speech language, a system of phonetic symbols which enabled the deaf to converse. Further work with the deaf continued and in 1873 he took up a professorship in vocal physiology at Boston University.

▸ IDEAS INTO PRACTICE

Bell's studies on sound waves were by now at an advanced stage, but his practical experiments were not. It was perhaps most important at this juncture then that he met the dexterous handy-man, Thomas Watson, who would help Bell's theoretical ideas become physical reality with the building of the Scotsman's designs. Watson also raised money to fund the work from the enthused parents of two of his deaf students, one of whom would eventually become his wife, Mabel Hubbard.

▸ SOUND WAVES

Bell's plans for voice communication were based around a single, simple concept. He believed that sound waves from the mouth could be converted into electrical current if the appropriate device could be created to make the conversion. Once this had been done, sending the current along a wire would be a relatively simple job, before positioning another device at the opposite end to reconvert the current into sound. After several taxing years working with Watson on perfecting the conversion device, he finally succeeded in producing a piece of equipment which would change the world at least as much as any other single creation before or since. Bell's telephone was patented in March 1876 and, although many disputes would follow concerning priority and copyright, the Scotsman took his place in history.

Still young, however, Bell used the money brought in by his invention, associated awards and his company AT & T, to fund the building of laboratories for further research. Just as Thomas Alba **Edison** (1847–1931) later improved the viability of Bell's phone through his carbon transmitter button, so Bell enhanced Edison's phonograph.

In addition, the Scottish inventor also worked on, among other things, sonar recognition, flying machines and the photophone, another sound transmitting device, this time employing light as its medium of transmission. Bell's work with the deaf continued too, including the development and improvement of teaching methods, as well as a period spent educating the now famous deaf and blind student, Helen Keller. Bell was also instrumental in the establishment of the international journal *Science*.

BELL'S LEGACY

For the man who is best remembered for an invention that allowed people to communicate over long distances using just the human voice, it is perhaps a little ironic that Alexander Graham Bell dedicated much of his life to working with the deaf.

When Bell was not teaching people who couldn't hear, he was inventing or dreaming up ideas for inventions. While the telephone was far and away the most successful of these, his passion meant that he became involved in a wide range of projects, many of them nothing to do with his principle invention, the telephone.

ANTOINE-HENRI BECQUEREL

1852–1908

CHRONOLOGY
• **1875** Becquerel begins research into various aspects of optics
• **1876** Takes up teaching post at the École Polytechnique in Paris
• **1888** Obtains his doctorate from the École • **1899** Elected to the French Academy of Sciences
• **1896** Discovers radioactivity • **1903** Awarded the Nobel Prize for Physics jointly with Marie and Pierre Curie

Wilhelm Conrad **Röntgen's** discovery of X-rays towards the end of 1895 would spark a flurry of investigations into the properties of the new phenomena. One of those stimulated into action by Röntgen's finding was the Frenchman Antoine-Henri Becquerel. Whilst attempting to further investigate X-rays in 1896, he chanced upon what is now known as radioactivity, opening up a whole new branch of scientific research.

▸ THE ENGINEER

Of course, by the time of the discovery for which he is remembered, Becquerel was already an established scientist, himself coming from a family noted for their scientific achievements. His grandfather and father were respected physicists in their own right and both had gone on to take up professorships at the French Museum of Natural History in physics, a feat which Antoine-Henri would also later mimic. In fact, his own son, Jean (1878–1953), would continue the family tradition following a similar path.

Investigating X-rays in 1896, Becquerel chanced upon the phenomenon of radioactivity

Like his immediate predecessors, Antoine-Henri was educated at the École Polytechnique in Paris, as well the School of Bridges and Highways, in engineering and the sciences. It was no surprise then that Becquerel eventually became chief engineer of the Department of Bridges and Highways. Nonetheless, he worked in parallel for much of his career in science and held many academic posts. Most notably, by 1895 Becquerel had become chair of physics at the École Polytechnique, and, again following in the footsteps of his father and grandfather, he had been honoured with membership of the prestigious Académie des Sciences in 1889, in recognition of his scientific endeavours. But Becquerel still, however, lacked his 'big break,' the achievement which would mark his place in the history of science. Almost certainly, he would have had little idea his innocuous researches into X-rays would take him there.

▶ AN INSPIRED HYPOTHESIS

The Frenchman's breakthrough began with a single hypothesis. Becquerel believed there was a possibility Röntgen's X-rays might also be responsible for the glowing or 'fluorescence' given off by some substances after being placed in sunlight. If this were the case, he deduced, the rays would make an impression on a covered photographic plate, passing through the protection as Röntgen had proven. Becquerel began to experiment on this basis. It just so happened one of the 'fluorescent' substances in which he had particularly expert knowledge was uranium, having undertaken prior work investigating its compounds. So, it was natural for Becquerel to use such a compound in his experiments and he found that after exposure to sunlight, the material did indeed mark the photographic plate.

It was when Becquerel packed his experimental equipment away, however, that his truly exciting discovery was made. After several days left in the dark, he took out the apparatus again, by now not fluorescent, and was amazed to discover the uranium compound, even after prolonged absence from sunlight, was still giving off sufficient radiation to impress upon the covered photographic plate (also stored next to the compound). He quickly realised this result was not down to X-rays but a new phenomenon for which he had no sound explanation, but which happened independent of any sunlight-induced luminescence. Further investigation would isolate the uranium as the cause of the 'radioactivity,' a name given to the occurrence not by Becquerel but Marie **Curie** (1867–1934), famous for her later researches into the subject. Becquerel was jointly awarded the Nobel Prize for physics with Marie Curie and her husband Pierre Curie (1859-1906) in 1903 for his work on radioactivity. The SI unit of radioactivity, the becquerel, is named after the Frenchman.

FURTHER ACHIEVEMENTS

The significance of Becquerel's chance discovery was not immediately realised and it was not until the Curies returned to the subject in 1898 that its potential and impact were first understood. One important observation which Becquerel himself later noted was the possibility of the use of radioactive materials in the medical world after he was burned by some radium in his pocket in 1901. Subsequent development would result in the use of the radiotherapy so common in cancer treatment today.

PAUL EHRLICH

1854–1915

CHRONOLOGY • 1882 Robert Koch discovers the tuberculosis bacillus • 1885 *Das Sauerstoff-Bedürfniss des Organismus* (*The Requirement of the Organism for Oxygen*) published • 1892 Ehrlich shows that mothers pass on antibodies through their breast milk • 1908 Receives the Nobel Prize for Physiology (jointly with Élie Metchnikoff) • 1909 Discovers an arsenic-based compound that combats syphilis

After the work of Edward **Jenner** (1749–1823) and Louis **Pasteur** (1822–95), the role and value of vaccinations were widely realised in the fight against disease. By the start of the twentieth century, however, there remained many untreatable fatal illnesses. Scientists began looking for alternative ways of conquering disease. One who was particularly successful and who, in the process, founded a new approach to the discovery of cures, was the German Paul Ehrlich.

▶ THE PRINCIPLE OF STAINING

Earlier in his career, Ehrlich had been deeply impressed by the development of a new finding involving the 'staining' of cells to highlight them when studied under the microscope. Some of these dyes only stained particular types of microorganism and Ehrlich was instrumental in creating a dye which illuminated the tuberculosis bacillus discovered by Robert Koch (1843–1910) in 1882. This was an important achievement in itself, becoming a technique widely used in the diagnosis of tuberculosis.

Ehrlich's 'magic bullet' became the cure for diseases such as tuberculosis and syphilis

▸ THE MAGIC BULLET

The principles behind staining remained central to most of the other work undertaken by Ehrlich during the rest of his career, and would provide the inspiration for the achievement for which he is most remembered. From around 1905 Ehrlich began to thoroughly research his hypothesis that if a dye could latch solely onto harmful bacteria (as he had proved with his previous work on tuberculosis), then perhaps other chemicals would behave in a similar way. Instead of illuminating the disease-causing microorganisms, however, he hoped they would kill them. The chemical which would become the basis for the proof of Ehrlich's theory was arsenic. This was an element potentially fatal to humans, but which in certain compounds he found could be used effectively to kill bacteria without a harmful number of side effects. Ehrlich at last completed the successful trial of the 'magic bullet' as a treatment for disease in 1909. An arsenic-based compound that he had been testing hunted out and killed the organism which caused syphilis. The following year he launched his treatment under the name Salvarsan, and it was hugely popular in combatting the disease, a widespread and unpleasant affliction often resulting in insanity and death. Moreover, the technique Ehrlich had employed was regarded as the foundation of chemotherapy, the treatment of disease by the use of synthetic compounds to locate and destroy the organisms causing an illness. It was an approach which would go on to have vital importance in combatting so many other diseases, most noticeably cancer-causing cells.

▸ A NOBEL PRIZE

In between his research on staining techniques and his cure for syphilis Ehrlich had jointly received a Nobel Prize for Physiology (in 1908) for a different discovery. From around 1889 to the turn of the century he was deeply involved with immunology and it was for this he received his award. He is often considered to be the founder of the modern approach to this area of science for his systematic and quantitative methods in attempting to understand it. He put forward theories on how the immune system worked and the role of antibodies. He also undertook a number of experiments designed to measure the increasing strength of the immune system in animals after repeated exposure to different types of disease-causing bacteria. This led to breakthroughs in the preparation of treatments for diphtheria and the development of techniques for measuring their effectiveness. Indeed, it was the later recognition of the limitation of these types of cures which would lead directly to Ehrlich's new approach to chemotherapy.

EHRLICH'S LEGACY

In our modern world with ready access to penicillin and other antibiotics, it is easy to forget the dreadful impact that diseases like smallpox and tuberculosis had on previous societies. Diseases which are now to all intents and purposes eradicated could spell a miserable death, even as little as fifty years ago. This is certainly the case with tuberculosis. In his obituary the London Times *paid tribute to Ehrlich's achievement in opening new doors to the unknown, acknowledging that, 'The whole world is in his debt.'*

NIKOLA TESLA

1856–1943

CHRONOLOGY • **1883** Tesla invents an induction motor • **1884** Arrives penniless in the United States • **1885** Westinghouse Electric buys the rights to Tesla's alternating current inventions • **1891** Invents the 'Tesla coil' • **c. 1899** Discovers terrestrial stationary waves • **1917** Tesla receives the Edison Medal

Few contemporaries of Thomas **Edison** (1847–1931) took him on and won, but a man who could make such a claim was one of the great American inventor's own former employees. Nikola Tesla, an eccentric electrical engineer, born in modern day Croatia but an emigrant to the U.S in 1884, was given work by Edison when he first landed in his new country. Contrasting personalities and conflicting ideas about electricity made the relationship a short one, sparking a bitter feud which would ultimately change the way the world received much of its power.

▶ A WAY TO TRANSPORT ELECTRICITY

The story begins earlier, though, back in Europe. The brilliant, eccentric and often troubled mind of Tesla was apparent from a young age. Although he did not come from an academic family background, there was a history of inventors in his ancestry and his father worked hard on developing Tesla's mental abilities. Despite interruptions to his childhood education due to frequent sickness and the severe trauma caused by the death of his older brother, Dane, Tesla progressed into higher education, taking up a place at the university of Graz in Austria.

The idea of the transmission of electricity without wires became a later interest for Tesla

Whilst at the university, Tesla was exposed to demonstrations of existing generators and electric motors and began to ponder better ways of creating and transporting electricity. He later came up with an idea involving a rotating magnetic field in an induction motor which would generate an 'alternating current' (now known as a.c.). Most electricity being created at the time for use in homes, offices and factories involved a direct current (d.c.) which had its limitations, particularly the cost of generating it, its difficulty in being transported over long distances and its need for a commutator. By contrast, Tesla would later prove his alternating current could travel safely, efficiently and cheaply over long distances. His invention of an induction motor, in line with his earlier ideas in 1883 was the first big step on that road. His next move was to sell it.

▸ AC OR DC?

Telsa decided to emigrate to America, arriving penniless but soon finding work by making use of his electrical engineering skills. Edison employed him, Edison fell out with him and Edison got rid of him within a year. But Edison's rival, George Westinghouse, wooed Tesla. In 1885 his company, Westinghouse Electric, bought the rights to Tesla's alternating current inventions and a war of electricity began. Edison and others believed in and, probably more importantly, had a financial interest in, direct current and wanted to make it a success. It was already the standard way of generating and supplying electricity. Westinghouse and Tesla believed their method was ultimately more adapted to the job and fought hard to promote it. In spite of Edison's attempts at damaging the reputation of alternating current by claiming it was unsafe (which Tesla would later refute by grand demonstrations lighting lamps using only his body, by allowing his a.c. current to flow through it), the greater benefits of alternating current were soon realised. With the subsequent invention of better transformers in its transportation, alternating current became the standard, with d.c. increasingly confined to only specialist applications. The trend continues to this day.

▸ THE TESLA COIL

In 1891, Tesla built on his knowledge to invent the 'Tesla coil' which was even more efficient at producing high frequency alternating current. It had many applications and still today is widely used in radio, television and electrical machinery. Using this and his turn of the century discovery of 'terrestrial stationary waves,' which basically meant the planet earth could be employed as an electrical conductor, he produced some spectacular demonstrations. Tesla generated self-made 'lightning-strikes' over a hundred feet long, and he once lit 200 lamps, unconnected by wires, stretched over 25 miles. Indeed, the idea of the widespread transmission of electricity without wires became a particular area of interest for Tesla in his latter years. The 'tesla', the SI unit of magnetic flux density, is named in honour of him.

FURTHER ACHIEVEMENTS

Tesla was also a prolific inventor. His inventions include: the telephone repeater, the rotating magnetic field principle, the polyphase alternating-current system, the induction motor, alternating-current power transmission, the Tesla coil transformer, wireless communication, radio, fluorescent lights, and more than 700 other patents.

SIR JOHN JOSEPH THOMSON

1856–1940

Towards the turn of the twentieth century, a time when many physicists believed most of the important discoveries in their subject had already been made, the Englishman John Joseph Thomson arrived and blew any such notions away.

Along with other scientific issues the nineteenth century had cleared up much of the confusion in atomic theory. Scientists believed, for example, that they now largely understood the properties and sizes of the atoms contained within elements; without question hydrogen was the smallest of all. So when 'J.J.' Thomson

announced the discovery of a particle one thousandth of the mass of the hydrogen atom, he rocked the scientific world.

▸ THE CATHODE RAY DEBATE

An early starter, Thomson attended classes in theoretical physics – a new subject at the time and one not on offer at all universities – at the University of Manchester when he was only fourteen. His most important discovery came while he held the chair of the now famous Cavendish Laboratory at Cambridge, a post he had taken over in 1884 and held until 1919. He had decided to investigate the properties of cath-

Instead of 'corpuscle', the tiny negatively charged particles were renamed 'electrons'

ode rays, now known to be a simple stream of electrons, but at the time the cause of widespread debate among scientists. The rays were visible, like normal light, but were quite clearly not normal light. Were they perhaps some form of X-ray? Most thought almost certainly not. To clear up the argument Thomson devised a series of experiments which would apply measurements to these cathode rays and clarify their nature.

▸ **MEASURING THE MASS OF PARTICLES**

The rays were created by passing an electric charge through an airless or gasless tube. By improving the vacuum in the tube Thomson quickly demonstrated that the rays could be deflected by electric and magnetic fields, a result which had not been observed before.

From this he concluded that the rays were made up of particles, not waves. Thomson then saw that the properties of the rays were negative in charge and didn't seem to be specific to any element; indeed, they were the same regardless of the gas used to transport the electric discharge, or the metal used at the cathode. Thomson devised a way of measuring the mass of the particles and found them to be consistently about a thousandth of the weight of the hydrogen atom. From these findings he concluded that cathode rays were simply made up of a jet of 'corpuscles,' and, more importantly,

that these corpuscles were present in all elements. He announced his discovery of the sub-atomic particle in April 1897 and in the process opened up a whole new branch of scientific research.

Thomson's conclusions were soon widely accepted but his terminology was not. Instead of the word 'corpuscle,' the tiny negatively charged particles were renamed 'electrons' and have been a fundamental part of the understanding of atomic science ever since.

▸ **THE CAVENDISH LABORATORY**

Thomson's position within the Cavendish Laboratory meant he also became involved in a range of other important physics projects, most notably involving the discovery of certain isotopes and aiding the development of the mass spectrograph. He was a superb teacher and leader, playing a vital part in the development of the reputation the Laboratory would gain as the world's leading authority in physics. Seven of his pupils went on to gain Nobel Prizes and, indeed, Thomson himself received the award for physics in 1906, as well as a knighthood in 1908: all from a man who had originally intended to go into engineering! Thomson had studied the sciences instead because he could not afford the fee to become an engineer's apprentice – his father had died in 1872. It was a strange quirk of fate for which physics would always be grateful.

FURTHER ACHIEVEMENTS

Although there are many claimants to the title of Father of modern physics, John Joseph Thomson's is probably as justified as anybody's. It was Thomson's discovery of the electron in 1897 which opened up a whole new way of looking at the world. Not only was matter composed of particles not even visible with a *modern electron microscope (as scientists from Democritus to Dalton had predicted) but it also appeared that those particles were composed of even smaller components themselves. Following Thomson, the discovery of these particles raised questions about the structure of matter that remain unanswered today.*

SIGMUND FREUD

1856–1939

CHRONOLOGY • 1886 Freud sets up his private clinic in Vienna • 1895 *Studies in Hysteria* published • 1896 Coins the phrase 'psychoanalysis' • 1899 *The Interpretation of Dreams* published • 1905 *Three Essays on the Theory of Sexuality* published • 1923 *The Ego and the ID* published

Sigmund Freud's popular impact remains profound even today. Yet for a scientist who changed the world, some critics would argue that his methods were at best unscientific and at worst downright reckless. Indeed, later thinkers in the fields of psychology and psychiatry have long since discredited many of his conclusions but still the Austrian's influence pervades. Whatever the rights or wrongs of his 'scientific' deductions, Sigmund Freud remains the benchmark by which others working in the same field must compare themselves and compete against.

▶ MEDICAL BEGINNINGS

Freud's entry into science was far less controversial. He began by studying medicine at the University of Vienna in 1873 and went on to take up a position at a hospital in the same city from 1882. It was time spent working with the French neurologist Jean-Martin Charcot (1825–93) in Paris from 1885, however, which set him on the path of his future career. Here he worked with patients suffering from hysteria and began to analyse the causes of their behaviour. Additional research with Josef Breuer back in Vienna during the early 1890s helped develop

'The interpretation of dreams is the royal road to the unconscious activities of the mind'

the basis for all of his future work, culminating in the publication of *Studies in Hysteria* in 1895.

▸ THE IDEA OF 'FREE ASSOCIATION'

In common with views generally held at the time, at the heart of Freud's conclusions was a belief that mental illness was normally a psychological rather than a physical brain disease. Once one accepted this premise then Freud's introduction of the idea of 'psychoanalysis' for diagnosing the causes of mental disorder (and indeed ultimately to explain all mental behaviour) was a logical one. One of the innovative tools he developed to aid in this was the idea of 'free association.' Rather than hypnotise people as was traditional, Freud advocated this method whereby patients enunciated thoughts or ideas which came into their consciousness without prior contemplation or analysis.

▸ DREAM THEORY

From this Freud believed he could make an insight into the 'unconscious' of a patient and, in particular, the 'repressed' thoughts and emotions (often related to past negative experiences) which their 'conscious' prevented from being articulated or enacted upon. For Freud, having a patient understand and acknowledge their repressed desires was a route to therapy and ultimately the treatment of a mental disorder. He also believed that dreams offered a major insight into repressed thoughts held in the unconscious mind. Indeed, his most prominent work which fully established his revolutionary approach – was entitled *The Interpretation of Dreams*, published in 1899.

While many critics could stomach, if not necessarily agree with, Freud's interpretations up until this point, he caused an outcry with his 1905 *Three Essays on the Theory of Sexuality*. His conclusions included the explanation that most repressed behaviour was in essence suppression of sexual impulses and, most shockingly, this activity began in infancy. It was here that he also introduced the now notorious concept of the Oedipus complex, a phrase used by Freud to describe feelings of sexual attraction of a child for its parent of the same sex, and hostility to the parent of the other sex. This phase , Freud claimed, speculatively at best, was one that all children passed through.

Gradually, however, Freud's analyses would gain credibility, if not necessarily with everyone, and certainly by the 1920s they had entered the popular consciousness on a global scale. He wrote many other texts including the 1923 *The Ego and the ID*. Freud effectively redefined the 'unconscious' as the 'ID,' an intangible collection of base impulses such as instincts and emotions present in the mind from birth. With experience, living and structure, aspects of the ID would gradually help formulate a person's 'ego.'

FREUD BY NAME, FREUDIAN BY NATURE

Freud's legacy remains as much in the tools of language that he has bestowed on the modern world as anything else. Terms he introduced or of which he altered the meaning to give them our now common understanding include: psychoanalysis, free association, the ID, the ego, neuroses, repression, the Oedipus complex and, of course, the Freudian slip. The structured, systematic approach he brought to analysing an inherently difficult-to-quantify subject also pervaded the work of his successors in the field.

HEINRICH RUDOLF HERTZ

1857–1894

• **1878** Hertz begins his PhD at the University of Berlin • **1880** receives his PhD • **1885** Appointed Professor of Physics at Karlsruhe Technical College • **1888** Discovers radio waves

Hertz came from a wealthy background and undertook his higher education initially at the University of Munich. In 1878, he began a PhD at the University of Berlin which he completed in 1880, still only twenty-three. By 1885, he was professor of physics at Karlsruhe Technical College, taking up a similar post at the University of Bonn in 1889. He had already completed his most memorable work by this point and within just five years he had died from blood poisoning.

▶ TESTING MAXWELL'S PREDICTIONS

The experiments for which Hertz became famous were undertaken in 1888. He had been developing them for about three years but had been considering them, at least theoretically, for a lot longer. His tutor whilst studying his PhD had suggested in 1879 experimental investigative work in the subject Hertz eventually examined, but it took the German several years to obtain the necessary equipment and facilities in order to conduct the tests. At their basis was an interest in James Clerk's Maxwell's vital prediction that there were almost certainly other forms of electromagnetic radiation similar in behaviour to

The name 'hertz' is familiar today to anyone who has ever switched on a radio

infrared, ultraviolet and normal visible light which lay out of known ranges at the time and would subsequently be discovered. Hertz hypothesised if this were true he could experimentally search for these waves through creating apparatus to detect certain electromagnetic radiation. He devised a machine which contained a circuit of electricity but with a gap in it which a spark would leap across when he chose to close the circuit. He reasoned if Maxwell's theory were true, appropriately sensitive equipment should pick up electromagnetic waves distributed by the spark and so he constructed the equivalent of an antenna. He placed this device across the room from the spark-creating circuit and, sure enough, the antenna detected waves. He called the waves 'Hertzian waves' but what he had in fact discovered, and they later became commonly known as, were radio waves.

▶ THE SPEED OF RADIO WAVES

Further experimental investigation showed these radio waves to have exactly the properties Maxwell had predicted. First and foremost, like other forms of electromagnetic radiation, they travelled at the speed of light. They could be reflected and refracted and made to vibrate in the manner of other waves. Indeed, as well as being significant for their importance as a newly found phenomena in their own right, Hertz's discovery

of radio waves and their properties was just as essential in that they conclusively and experimentally proved Maxwell had been correct when suggesting light, and additionally heat, waves were all forms of electromagnetic radiation.

▶ A USELESS DISCOVERY?

The German did not immediately see himself, however, the true significance of his experimental results beyond merely proving Maxwell's theories had been correct. When asked what physical application could be made of his discovery, Hertz answered, 'It's of no use whatsoever. This is just an experiment that proves Maestro Maxwell was right. We just have these mysterious electromagnetic waves that we cannot see with the naked eye. But they are there.' Others did not accept such a conclusion as easily, though, and when Hertz published the methods and results of his experiments they began looking at ways of exploiting these radio waves.

▶ MARCONI'S DEVELOPMENT

Unfortunately, Hertz did not live long enough to see the practical use one of those inspired by his essay, Guglielmo **Marconi** (1874-1937), would make of his discovery. The Irish-American transimitted radio signals over increasing lengths towards the end of the century, and succeeded in sending a signal across the Atlantic by 1901.

THE LEGACY OF HERTZ

His surname is one with which the world is familiar even today. In honour of Heinrich Rudolf Hertz's achievements, the SI unit of frequency, the hertz, was named after him. Indeed, the fact that people most commonly encounter his name when tuning in their radio is indicative of his accomplishment in itself. What is perhaps less well known is that the German physicist made his breakthrough at a very young age. By the time he was just thirty-six, he was dead.

145

MAX PLANCK

1858–1947

CHRONOLOGY • **1892** Planck appointed Professor of Theoretical Physics at the University of Berlin • **1 January 1900** First public enunciation of quantum theory • **1900** 'On the Theory of the Law of Energy Distribution in the Continuous Spectrum' published • **1918** Planck awarded the Nobel Prize for Physics

When did the 'modern' scientific era actually begin? Throughout the nineteenth century, there had been many advances in all aspects of science which could arguably be seen to have launched a new foundation. But for physics at least, the answer is simple: 1 January 1900. That was the day the German Max Planck gave the first public enunciation, albeit to his son, of quantum theory. It was a notion which completely abandoned assumptions made in classical physics and it founded a whole new age.

▶ QUANTUM THEORY

To a degree, Planck stumbled across the concept that would change the scientific world through an element of chance. He had been undertaking theoretical work in thermodynamics and it was his search for a hypothetical answer to an inexplicable problem in physics at that time which led to a solution reflected in reality. The German, like many other scientists before him, had been considering formulae for the radiation released by a body at high temperature. He knew it should be expressible as a combination of wavelength frequency and temperature, but the 'irregular'

Planck stumbled across the concept that would change the scientific world by chance

behaviour of hot bodies made a consistent prediction difficult. For a theoretically perfect form of such a matter known as a 'black body', therefore, physicists could not predict the radiation it would emit in a neat scientific formula. Earlier scientists had found expressions which were in line with the behaviour of hot bodies at high frequencies, and others found an entirely different equation to show their nature at low frequencies. But none could be found which fitted all frequencies and obeyed the laws of classical physics simultaneously.

▶ PLANCK'S CONSTANT

Not that such a conundrum bothered Max Planck. Instead, he resolved to find a theoretical formula which would work mathematically, even if it did not reflect known physical laws. The answer he soon found was a relatively simple one: the energy emitted could be expressed as a straightforward multiplication of frequency by a constant which became known as 'Planck's constant' (6.6256×10^{-34} Js). But this only worked with whole number multiples (e.g. 1, 2, 3, etc.) which meant that for the formula to have any practical use at all, one had to accept the radical assumption that energy was only released in distinct non-divisible 'chunks', known as quanta, or for a single chunk of energy, a quantum. Up

until that point it had been assumed that energy was emitted in a continuous stream, so the idea it could only be released in quanta seemed ridiculous. It completely contradicted classical physics. But Planck's explanation fitted the behaviour of radiation being released from hot bodies. Moreover, the individual quanta of energy were so small that when emitted at the everyday, large levels observed in nature, it seemed logical energy could appear to be flowing in a continuous stream.

▶ BIRTH OF A THEORY

In this way classical physics was cast into doubt and quantum theory was born. Planck announced his results to the wider public in his 1900 paper 'On the Theory of the Law of Energy Distribution in the Continuous Spectrum'. It naturally caused a stir, but when Albert **Einstein** was able to explain the 'photoelectric' effect in 1905 by applying Planck's theory, and likewise Niels Bohr in his explanation of atomic structure in 1913, the notion suddenly did not seem so ridiculous. The abstract idea really could explain the behaviour of physical phenomena and consequently Planck was quickly elevated to the status of Germany's most prominent scientist. He was awarded the Nobel Prize for Physics for his breakthrough in 1918.

FROM CLASSICAL PHYSICS TO QUANTUM MECHANICS

The classical physics of Newton and Galileo provides us with laws capable of explaining the ordinary, everyday world around us. However, experiments conducted early in the twentieth century began to produce results that could not be explained by classical physics. One example was the discovery that if the electrons of an atom orbited the nucleus in the way classical

physics predicted, they would spiral down into the nucleus within a very short time, and the atom would cease to exist. As this clearly was not the case, it became clear that another way of dealing with atomic and subatomic particles was required. This discovery, allied to Planck's quanta theory of energy, led to the development of quantum mechanics.

LEO BAEKELAND

1863–1944

CHRONOLOGY • **1863** Baekeland born in Ghent, Holland • **1887** Appointed Professor of Physics and Chemistry at Bruges • **1888** Returns to Ghent as Assistant Professor of Chemistry. • **1889** Frustrated by academic life, Baekeland settles in America while on honeymoon • **1899** Baekeland's first company, a manufacturer of photographic paper, is bought by the Kodak Company for $1,000,000 • **1909** Sets up General Bakelite Company (GBC) • **1939** GBC becomes a subsidiary of the Union Carbide and Carbon Corporation

Baekeland worked at first as a photographic chemist and in 1891 he opened his own consulting laboratory. In 1893 he began to manufacture a photographic paper, which he called Velox, and six years later his company was bought out by the Kodak Corporation for one million dollars. Now financially independent, Baekeland returned to Europe to study at the Technical Institute at Charlottenburg.

Throughout the nineteenth and into the twentieth century, the United States of America became increasingly influential in many arenas. Its impact on the world of science was no exception. The country was also renowned for its encouragement of entrepreneurship – anybody who could combine a talent for scientific originality with a flair for business not only stood a good chance of discovering something which could change the world, but could also become incredibly wealthy off the back of it! The Belgian born

Baekeland chased the American dream and was rewarded very handsomely indeed

immigrant Leo Hendrick Baekeland was one such chemist. He chased the American Dream and was rewarded very handsomely indeed.

▸ MAKING PLASTIC

The creation for which Baekeland is best remembered and certainly the one which would have the most impact on the modern world was his development of the first widely useful synthetic plastic, a product which would find many future applications. His journey of discovery began in 1905. Baekeland began work on new chemical experiments after returning to the United States from a spell of study in Europe at the Charlottenburg Technical Institute. He had decided to try and produce a synthetic version of shellac: thin plates created by melting naturally occurring crimson-coloured lac. For this purpose he chose to work with a product of formaldehyde and phenol discovered over thirty years earlier but never commercially developed. The scientist quickly found his search for synthetic shellac to be a lost cause. Nevertheless, he became interested in the properties of the materials he was working with. After further experimentation he found that if he brought the formaldehyde and phenol together under a combination of high temperature and pressure, they produced a stiff, hardwearing resin. He had, in fact, produced the first plastic by thermosetting and was quick to realise the potential for its wide-ranging commer-

cial use. In 1909 he launched his product under the name Bakelite and it became hugely popular both industrially and domestically. The family of plastics quickly grew and the world was never quite the same again.

▸ PHOTOGRAPHIC PAPER

Although the launch of Bakelite was commercially very successful, Baekeland had made his fortune long before this through another extremely useful invention which took advantage of his prior chemical knowledge. The Belgian had moved to the United States in 1889 after visiting during his honeymoon. Prior to this time, however, he had held academic posts in his home country, in both physics and chemistry. On arrival in America he decided to abandon formal study and took a job in a photographic laboratory. Bringing his knowledge of science to the fore, he invented a special type of photographic paper which could be developed under artificial light. Baekeland then left regular employment to set up his own company. In 1893 he launched his new creation under the brand name Velox. It was the first photographic paper in the world to be both successful and widely used, and became instrumental in the growth of the photographic industry. This was reflected in the one million US dollars paid for Baekeland's firm just six years later by the Kodak Corporation, a huge sum in Baekeland's time and not a small amount today.

THE INFLUENECE OF BAEKELAND

Baekeland made not just one important contribution to the modern world through his chemistry, but two. The United States had served the fulfilment of his dreams well and he had served his adopted country likewise. This was acknowledged not just in the money Baekeland made but also in the many awards

and honours which were bestowed upon him for his efforts, including the presidency of the American Chemical Society in 1924.

The synthetic bakelite was in effect the first plastic, one of the most common materials in everyday use in the modern world.

THOMAS HUNT MORGAN

1866–1945

The 'rediscovery' in 1900 of the laws of inheritance first observed by Johann Gregor **Mendel** (1822–84) excited many biologists who at last believed they had found an explanation for hereditary traits, and quite possibly the mechanism to underpin Darwin's theories. One scientist who initially remained unmoved and unconvinced by them, however, was the American Thomas Hunt Morgan.

▸ **HEREDITY AND THE CELL**

After earlier work in embryology, Morgan dedicated most of the period between 1904–28, while he was Professor of Zoology at Columbia University, to clarifying how hereditary processes worked. He started out with Mendel's laws of segregation and independent assortment and began to critically assess them. It was not so much that he doubted the outcome of hereditary traits predicted by the laws; experimental evidence often seemed to back up the mathematical forecasts for characteristics present in descendants that Mendel had suggested. Morgan felt it to be more that they could not accurately reflect the process of arriving at the end result, in

Morgan began breeding the fruit fly in 1908, work which was to make him famous

particular the law of independent assortment. The reason the American felt this way was because it had been separately established that chromosomes – long thread-like matter present in the nucleus of a cell which grew and divided during cell splitting – clearly played an important part in inheritance. Yet there were far fewer chromosomes in living things than there were 'units of heredity' (renamed 'genes' in 1909 by the Dane Wilhelm Johannsen). To Morgan, this meant that groups of genes had to be present on a single chromosome. This would implicitly invalidate Mendel's law of independent assortment (which dictated that hereditary traits caused by genes would occur in all possible mathematical combinations in a series of descendents, independent of each other).

▶ BREEDING THE FRUIT FLY

From 1908 Hunt began to investigate, breeding the fruit fly, which has just four pairs of chromosomes. It is for this work that he eventually became famous. Early into his studies, he observed a mutant white-eyed male fly which he extracted for breeding with ordinary red-eyed females. Over subsequent generations of interbred offspring, the white-eyed trait returned in some descendants, all of which again turned out to be males. It was exactly the link Morgan had been looking for. Clearly, certain genetic traits were not occurring independently of each other but were actually being passed on in groups. At this point, though, Morgan realised that rather than invali-

date Mendel's law of independent assortment, a simple adjustment was all that was required to unite it with his belief in the importance of chromosomes and produce an all- encompassing, proven thesis. He suggested that the principle of independent assortment did apply, but only to genes found on different chromosomes. For those on the same chromosome, linked traits would be passed on, usually a sex-related factor with other specific features (for example, the male sex and the white-eyed characteristic in the fruit fly). Otherwise, Morgan now accepted Mendel's laws.

▶ THE CHROMOSOME MAP

The results of his work had convinced Morgan that genes were arranged on chromosomes in a linear manner and could actually be 'mapped'. Further testing showed that the linked traits that Morgan had previously observed could occasionally be broken during the exchange of genes that occurred between pairs of chromosomes during the process of cell division. The American suggested that the nearer on the chromosome the genes were located to each other, the less likely the linkages were to be broken. Thus, by measuring the occurrence of breakages he could work out the position of the genes along the chromosome. Consequently, in 1911, he produced his first chromosome map showing the position of five genes which were linked to gender characteristics. Just over a decade later Morgan, together with other scientists, had mapped two thousand genes on the chromosomes of the fruit fly.

FURTHER ACHIEVEMENTS

Of Morgan's many books two in particular deserve special attention: The Mechanism of Mendelian Heredity *(1915) and* The Theory of the Gene *(1926). They laid the basis for understanding Mendel's observations and, along with* *work by later geneticists, helped to provide the microscopic science required to reinforce Charles Darwin's conclusions. In 1933 Morgan received the Nobel Prize for Physiology.*

MARIE CURIE

1867–1934

CHRONOLOGY • 1893 Curie graduates in physics from the Sorbonne. She is top of her class • 1898 Discovers the elements polonium and radium • 1903 Awarded the Nobel Prize for Physics (jointly with her husband Pierre Curie and Henry Becquerel) • 1910 *Treatise on Radioactivity* published • 1911 Awarded the Nobel Prize for Chemistry

Quite aside from her practical achievements, Marie Curie's is also important in the history of science for the pioneering role she played in opening up the subject to other women. She was arguably the first globally renowned and accepted female scientist and as such forged a path for all those of her sex who followed her. Her scientific discoveries in themselves were vital to understanding the new phenomenon of radioactivity. This was reflected in the fact that she was awarded not one, but two Nobel Prizes.

The majority of Curie's scientific work would take place in France, where she spent most of her life from 1891. Her country of birth, however, was Poland where she was born under the name Marya Sklodowska. Despite both her parents' status as teachers she grew up in relative poverty there. This was further accentuated when she was forced to move to Paris in order to obtain a higher education in physics, a level of study women were unable to undertake in her home country at the time. She graduated and shortly afterwards met her future husband, Pierre Curie (1859–1906), at the Sorbonne, where she studied and he worked.

Even today, Marie Curie's notebooks of her studies remain too radioactive to handle

He was a respected physicist in his own right and it is no surprise that the two began working together in 1895, not long after they married.

▶ IN THE FOOTSTEPS OF BECQUEREL

The stimulus for the couple's later achievements would come initially from Marie's hunt for an area of research to undertake for her postgraduate studies. Encouraged by Pierre, she decided to further investigate the exciting, new discovery of radioactivity made by Henry **Becquerel** (1852–1908) in 1896. Curie's investigations into better understanding the properties of the phenomenon soon yielded results. Becquerel had proved that uranium was radioactive. Curie, wanting to find out which other elements were, quickly discovered that thorium shared similar traits. She went on to conclusively prove that radioactivity was an intrinsic atomic property of the element in question – uranium for example – and not a condition caused by other outside factors.

Curie's next achievement was to actually discover two new elements in 1898 through her researches, which she called polonium and radium, both highly radioactive, especially the latter. She had tracked down these elements after realising uranium ore had a greater level of radiation than pure uranium, thereby correctly deducing that the ore must have contained other more radioactive, hidden elements. After these discoveries, Curie sought to obtain large enough quantities of the new substances to further understand their properties. Unfortunately, because of the minute amounts in which radium in particular was present in uranium ore, this meant she, along with her husband, had to wade through tonnes of the stuff for several years just to obtain a tenth of a gram by 1902. This, at least, allowed the calculation of the atomic weight of the new element to be made, as well as other work on its properties.

▶ A QUESTION UNANSWERED

There was one question the Curies never fully got to the bottom of, however. What exactly was the radiation which came from these elements? Ernest **Rutherford** (1871–1937) would take the credit for the answer to this question with his explanation of 'alpha,' 'beta,' and later 'gamma,' rays, but Marie did observe that the radiation was made up of at least two types of rays with distinct individual properties.

Sadly, Marie Curie would eventually die from leukaemia, which is thought to have been caused by her long exposure to radiation. At the time of her work with radioactive elements, the risks associated with radiation were not known and so no precautions were taken. Even to this day, her notebooks from her period of radioactive study remain too dangerous to examine.

MARIE CURIE'S LEGACY

Marie Curie would go on to be elected to her husband's former post of Professor of Physics at the Sorbonne, and become the institution's first female professor in the process. Her achievements in this position included the establishment of a research laboratory for radioactivity in 1912. The laboratory would go on to become world-renowned for its contribution to physics. This was due in no small part to a gift bestowed on Curie by the United States in 1921 which greatly facilitated the centre's work: a gram of the rare radium.

She received her second Nobel Prize in 1911 (her first was awarded with Pierre jointly in 1903), this time in chemistry, in recognition of her discovery of polonium and radium.

ERNEST RUTHERFORD

1871–1937

CHRONOLOGY • 1902 Rutherford establishes new branch of physics with Frederick Soddy: radioactivity • 1908 Awarded Nobel Prize for Chemistry • 1911 Establishes nuclear theory of the atom • 1914 Becomes Sir Ernest Rutherford • 1919 Develops proton accelerators (atom smashers)

After the discovery by Antoine-Henri **Becquerel** (1852–1908) of radioactivity in 1896, a number of scientists became responsible for a deeper understanding of the new phenomenon, including of course Marie **Curie** (1867–1934). However, the person who perhaps did most to bring a full understanding of radioactivity to the world, and greatly develop nuclear physics in general, was Ernest Rutherford.

▸ THE CAVENDISH LABORATORY

The New Zealand-born scientist won a scholar-ship to the Cavendish Laboratory in Cambridge in 1895, working under the eminent J. J. **Thomson** (1856–1940). He went on to become a professor at the McGill University in Montreal in 1898. Here he put forward his observation that radioactive elements gave off at least two types of ray with distinct properties. These he named 'alpha' and 'beta' rays.

▸ GAMMA RAY THEORY

In 1900 he confirmed the existence of a third type of ray, the 'gamma' ray, which was distinct in that it remained unaffected by a magnetic force, while

The discovery that atoms could simply decay away from an element was remarkable

alpha and beta rays were both deflected in different directions by such an influence.

▸ DISCOVERING THE HALF-LIFE

It was also in Montreal that Rutherford met the British chemist Frederick **Soddy** (1877–1956). Between 1901 and 1903, the two collaborated on a series of experiments related to radioactivity and came to some startling conclusions. They showed how, over a period of time, half of the atoms of a radioactive substance could disintegrate through 'emanation' of a radioactive gas, leaving 'half-life' matter behind. The notion that atoms could simply decay away from an element was quite remarkable. Moreover, during the process the substance spontaneously transmuted into other elements – a revolutionary finding.

After the collaboration with Soddy ended, Rutherford went on to examine alpha rays more closely. He soon proved through experimental results that they were simply helium atoms missing two electrons (beta rays were later shown to be made up of electrons and gamma rays, actually short X-rays). During this period he moved back to England, to take up a post at Manchester University. Here he worked with Hans Wilhelm Geiger (1882–1945) to develop the Geiger counter in 1908. This device measured radiation and was used in Rutherford's work on identifying the make-up of alpha rays. He went on to use it even more significantly in his next major advance.

▸ BOUNCING ALPHA PARTICLES

In 1910 Rutherford had proposed that Geiger and another assistant should undertake work to examine the results of directing a stream of alpha particles at a piece of platinum foil. While most passed through and were only slightly deflected, about one in eight thousand bounced back virtually from where they had come! Rutherford was astonished, describing it later as 'quite the most incredible event that has ever happened to me in my life. It was almost...as if you had fired a fifteen-inch shell at a piece of tissue paper and it came back and hit you.'

He didn't let it fox him. In 1911 he put forward the correct conclusion: the reason for the rate of deflection was because atoms contained a minute nucleus which bore most of the weight, while the rest of the atom was largely 'empty space' in which electrons orbited the nucleus much as planets did the Sun. The reason that the one in eight-thousand alpha particles bounced back was because they were striking the positively charged nucleus of an atom, whereas the rest simply passed through the spacious part. It was a vital discovery on the path to understanding the construction of the atom and would greatly aid Niels **Bohr** in his related revelations of 1913.

FURTHER ACHIEVEMENTS

During the First World War Rutherford served in the British Admiralty. Afterwards, in 1919, he was appointed to the chair of the Cavendish Laboratory at Cambridge. In the same year he made his final major discovery. Working in collaboration with other scientists, he found a method by which he could artificially disintegrate an atom by inducing a collision with an alpha particle. Essentially, what we now know as protons could be forced out of the nucleus by this smash. In the process the atomic make up of the substance changed, thereby transforming it from one element to another.

In this first instance he transmuted nitrogen into oxygen (and hydrogen), but went on to repeat the process with other elements.

THE WRIGHT BROTHERS

WILBUR 1867–1912 · ORVILLE 1871–1948

• **4 June 1783** First public demonstration of the Montgolfiers' hot air balloon • **1896** German Otto Lilienthal dies in a glider accident. This marks the beginning of the Wrights' serious interest in flight • **17 December 1903** Orville makes the first successful powered, sustained and controlled flight in heavier-than-air machine 'The Flyer' • **1908** The Wrights reveal their improved machines to the public • **8 August 1908** First public flight in Europe at Le Mans in France

The Wright brothers' interest had not always been in flying. Their first joint project, after leaving school, was starting and publishing their own newspaper. They later moved into bicycles, opening their own shop building and selling them. So it was only from 1896 their curiosities turned to aircraft. Others had already built gliders and it was the death in a crash of one of the early pioneers of these machines, the German Otto Lilienthal (1848–96) which initially drew their attention to the subject. Lilienthal had made a number of advances in understanding aerodynamics, so the brothers began by studying his and other inventors' progress to date.

▶ SYSTEMS OF CONTROL

The Wrights quickly realised whilst much of the focus in the pioneering of gliders had been on making the craft stable, this had been at the expense of any real kind of control. So they systematically began experimentation on control

The Wrights quickly realised that gliders, though stable, lacked control

mechanisms which would eventually include ground-breaking mechanisms for the twisting of their aeroplane's wings, inspired after observing the flight of birds when searching for aerodynamic clues from nature.

▶ MASTERING THE AERODYNAMICS OF THE GLIDER

The Americans' research also involved the building and testing of unmanned gliders which they refined after each experiment. By 1900 they had built their own manned glider. They further improved their testing with the construction of a wind tunnel to aid their research in 1901 and by the following year they had made enough findings to build the most successful glider anywhere in the world up to that time. It was almost inevitable, therefore, that once they had mastered the aerodynamics of aircraft construction, it would only be a matter of time before they took the next logical step of adding motors to their invention. Before they could do this, however, they needed a method of converting the electrical power of the motor into forward thrust. They found this method in the propellor.

The Wrights tested numerous prototypes of propellor, which they conceived as moveable wings rotating around fixed axes, until they were finally satisfied. They then built their own 12 horsepower engine which powered two propellors and fitted it to their biplane, The Flyer.

The first ever successful manned flight in a powered aircraft took place, like all of their other tests, at Kitty Hawk, North Carolina on 17 December 1903. Wilbur had made an unsuccessful launch a few days earlier, but it was Orville who became the first to go airborne in a controlled manner on this date, initially covering just 120ft. By the end of the day, they had both made successful, longer flights.

▶ AIRCRAFT ON DISPLAY

Over the next few years the brothers worked on improved versions of their early aircraft, waiting until they had much more reliable models in place before demonstrating their advances to the public. Indeed, it was not until 1908 that they revealed their machines: Wilbur gave a display in France and Orville in Virginia, USA, within a few days of each other. They went down well, for within a year the brothers had received the backing to build their planes commercially and with great success in both the USA and Europe. Sadly, Wilbur caught typhoid fever in 1912 and died, leaving Orville to inherit the business and a sizeable fortune when he sold it in 1915.

FURTHER ACHIEVEMENTS

Although the human race had been airborne for over a hundred years before Orville Wright and his elder brother Wilbur (1867–1912) came along, it had struggled to make the next logical step in the progress of flight. For thousands of years mankind had dreamed of flying until the Montgolfiers had fulfilled that fantasy with their hot air balloon in 1783. This had taken advantage of the fact that some gases were 'lighter-than-air' and as such would propel a
suitable device, such as a balloon, upwards. For the century afterwards, however, inventors had fantasised about achieving another seemingly impossible dream; powered flight with 'heavier-than-air' machines. Many tried to create such devices during the nineteenth century, and many failed. Then the Wright brothers arrived with a completely different approach and by 1903 had invented and tested the world's first motor-powered aeroplane.

GUGLIELMO MARCONI

1874–1937

CHRONOLOGY • **1896** Marconi files first patent for radio equipment • **1897** Sends radio signal almost 9 miles across the Bristol Channel • **1898** Sets up the Wireless Telegraph and Signal Company, Ltd • **1909** Awarded the Nobel Prize for Physics jointly with Ferdinand Braun • **1918** Sends first radio message from England to Australia

Some people make remarkable discoveries and others take advantage of them. Guglielmo Marconi was not in essence a scientist but instead a brilliant collaborator and manipulator of other scientists' findings. The most important of these was Heinrich Rudolf **Hertz's** (1857–94) breakthrough (the discovery of radio waves in 1888), with ramifications even he could not have realised. One person who did quickly realise their significance, however, was the Irish-Italian Marconi. And it was he, not Hertz, who ended up with a Nobel Prize for Physics in 1909.

▶ FIRST EXPERIMENTS

As well as being more interested in the scientific implications of his findings than the practical or commercial use of them, the main reason Hertz had not been able to take much advantage of his discovery was because he had died shortly afterwards in 1894. At exactly that time, Marconi was beginning to undertake his first experiments with radio waves after having his interest stimulated when learning about the work of Hertz and others during his studies in Bologna and Livorno. Both his father and mother had been wealthy even before they had married, so Marconi had

The 2,000-mile radio transmission across the Atlantic in 1901 convinced the last doubters

the advantage of a large family estate and limited financial worries in which to begin his experiments.

▸ WORK IN LONDON

Indeed, the size of the estate became important as he made his first developments. It was a mile and a half long so he had the benefit of a large private testing ground across which to send his radio waves. By 1895 he had made sophisticated enough equipment to emit and receive waves the entire length of his grounds. Whilst Marconi was convinced of the importance of his test results, few others in Italy were, so he set off for London, England instead in an attempt to stimulate backing for his work there.

He was far more successful. Government, military and postal departments quickly became interested in the potential uses of the new technology and within a few more years, enthusiasm was widespread. This came largely after his first overseas radio transmission between Britain and France in 1899, after which he was brought to the attention of the public. During the intervening period, he had also been involved in the setting up of a company to promote his develop-

ments, renamed in acknowledgement of his centrality to it as the Marconi Wireless Company Limited in 1900. By that time he had also held successful radio transmission trials on floating vessels on behalf of the British navy, after which naturally followed further converts to his cause.

▸ ACROSS THE ATLANTIC

But the event which made Marconi world famous, and the one which silenced many of the doubters about the practical and scientific uses of his equipment who still persisted, was the two thousand mile transmission of Morse code across the Atlantic in 1901. Many had thought this an impossible task, not least because the curvature of the earth would prevent it, but Marconi had been convinced it was possible and was duly rewarded for his perseverance.

Both before and after this time, the Irish-Italian took out many patents for the equipment he had successfully developed and added to. Indeed, his life remained absorbed by improvements to radio technology virtually up until his death, perhaps most notably making advances in short-wave technology to set up an international network for radio broadcasts by 1927.

MARCONI AND THE TITANIC

'Those who had been saved, had been saved through one man, Mr. Marconi.... and his marvellous invention' (The Right Hon. Herbert Samuel, the Postmaster General April 18, 1912).

Although there was initially a great deal of scepticism regarding the usefulness of Marconi's invention, once he had demonstrated its potential, it quickly became something of a craze. One particular area it was warmly

welcomed was shipping; before long, every ocean-going liner had its 'Marconi man'. One reason as many survivors were picked up from the Titanic when she went down was her Mrconi men, Jack Phillips and Harold Bride, who stayed on the ship transmitting a distress call until the power failed, and the radio room was awash. Harold Bride survived the ordeal, and later joined the Royal Navy; Jack Phillips went down with his ship.

FREDERICK SODDY

1877–1956

CHRONOLOGY • **1898** Graduates from Merton College, Oxford with first class honours in chemistry • **1901–03** With Ernest Rutherford, discovers the phenomenon of atomic disintegration • **1904–14** Lecturer in Physical Chemistry at Glasgow University • **1919** Appointed Dr. Lees Professor of Chemistry at Oxford University

While perhaps not as pre-eminent as his contemporary Ernest Rutherford (1871-1937), Frederick Soddy was still a significant influence in understanding the behaviour and implications of radioactivity, particularly of radioactive decay. The English chemist's first breakthrough came after working with Rutherford between 1901–03 while at the McGill University in Montreal. Moreover, the later solitary work which this led to for Soddy would help scientists understand an important aspect of the atomic design of elements.

▶ DECAYING ATOMS

The work Soddy undertook with Rutherford came shortly after he took up his position as demonstrator of chemical experiments in Montreal. Their collaboration yielded radical results, most notably the revolutionary idea that elements could disintegrate into a 'half-life' by 'emanating' atoms spontaneously. In the process of this decay, matter could transmute into various other elements, an astonishing conclusion.

These findings in themselves established Soddy's reputation, but his later results in related work were just as important. He turned his atten-

Soddy's discovery of isotopes cleared away a good deal of chemical confusion

tion to the array of apparent 'new' elements which had been discovered during the transmutation caused by radioactive decay. The problem scientists were facing was that there had been so many recent discoveries there were clearly were not enough spaces in which to fit them in the periodic table. Yet each element definitely had a distinct atomic weight and specific time period ascribed to it before it reached its 'half-life' state. Meanwhile, other scientists were trying to produce these new elements artificially by breaking-down closely related elements but were inexplicably failing.

▸ ISOTOPES

Soddy's 1913 solution was another simple, yet incredible, one. He maintained that even though the atomic weights and half-lives of the 'new' elements were different, they otherwise shared identical chemical properties with known existing elements and were therefore variations of the same element. So, for example, the 'new' elements thorium C and radium D, which had different atomic masses and half-lives from each other and from the element lead, were all chemically the same and were therefore all simply lead. This also explained why scientists had not been able to break down artificially the closely related elements, as expected, because the substances

they were starting with and seeking to produce were chemically identical! Soddy named the variations 'isotopes' and in a stroke cleared up the confusion which had previously surrounded the 'new' elements. It would take the later discovery by James **Chadwick** (1891-1974) of the neutral, but weight bearing, neutron to fully explain how the differences in atomic mass were possible whilst still retaining the same atomic number, but Soddy's explanation offered enough for science to make some sense of the new order in the meantime.

▸ RADIOACTIVE DISPLACEMENT

In the same year as his isotopic explanation, Soddy also expressed the radioactive displacement law. This stated that when an alpha particle is emitted from a decaying substance, its atomic number is reduced by two, and its atomic weight by four (explicable by the fact that an alpha particle is simply a helium nucleus with corresponding atomic figures). Likewise when a beta particle (a negatively-charged electron) is emitted, the atomic number is increased by one.

Soddy was awarded the Nobel Prize for chemistry in 1921 for his work on isotopes, after which his interest moved into other academic disciplines and he ceased active involvement in chemical research.

SODDY'S LEGACY

It was during his time at Glasgow, between 1904 to 1914 that Soddy did much of his most important work in chemistry, including his formulation of the 'Displacement Law', which stated that emission of an alpha particle from an element causes that element to move back two places in the Periodic Table. He also

formulated the concept of isotopes during thie period, the realisation that elements can exist in two different states, with different atomic weights, while remaining chemically identical. Soddy devoted much of his later time to fields other than chemistry, evolving theories which were never widely accepted.

ALBERT EINSTEIN

1879–1955

• **1902** Einstein begins work at Swiss patent office • **1905** Publishes three seminal papers on theoretical physics, including the 'special' theory of relativity • **1916** Proposes general theory of relativity; is proved correct three years later • **1922** Wins Nobel Prize in Physics • **1933** Emigrates to Princeton, N.J. • **1939** Urges Franklin D. Roosevelt to develop the atom bomb • **18 April 1955** Dies in his sleep

O f the essays written by Einstein in 1905, arguably the most influential was his enunciation of a 'special' theory of relativity, which advanced the idea that the laws of physics are actually identical to different spectators, regardless of their position, as long as they are moving at a constant speed in relation to each other. Above all, the speed of light is constant. It is simply that the classical laws of mechanics appear to be obeyed in our normal lives because the speeds involved are insignificant.

▶ THE SPEED OF LIGHT

But the implications of this principle if the observers are moving at very different speeds are bizarre and normal indicators of velocity such as distance and time become warped. Indeed, absolute space and time do not exist. Therefore if a person were theoretically to travel in a vehicle in space close to the speed of light, everything would look normal to them but another person standing on earth waiting for them to return would notice something very unusual. The space ship would appear to be getting shorter in the

'Science without religion is lame; religion without science is blind'

direction of travel. Moreover, whilst time would continue as 'normal' on earth, a watch telling the time in the ship would be going slower from the earth's perspective even though it would seem correct to the traveller (because the faster an object is moving the slower time moves). This difference would only become apparent when the vessel returned to earth and clocks were compared. If the observer on earth were able to measure the mass of the ship as it moved, he would also notice it getting heavier too. Ultimately nothing could move faster than or equal to the speed of light because at that point it would have infinite mass, no length, and time would stand still!

▶ A GENERAL THEORY OF RELATIVITY

From 1907 to 1915, Einstein developed his special theory into a 'general' theory of relativity which included equating accelerating forces and gravitational forces. Implications of this extension of his special theory suggested light rays would be bent by gravitational attraction and electromagnetic radiation wavelengths would be increased under gravity. Moreover mass, and the resultant gravity, warps space and time, which would otherwise be 'flat', into curved paths which other masses (eg, the moons of planets) caught within the field of the distortion follow.

Amazingly, Einstein's predictions for special and general relativity were gradually proven by experimental evidence. The most celebrated of these was the measurement taken during a solar eclipse in 1919 which proved the sun's gravitational field really did bend the light emitted from stars behind it on its way to earth. It was the verification which led to Einstein's world fame and wide acceptance of his new definition of physics.

Einstein spent much of the rest of his life trying to create a unified theory of electromagnetic, gravitational and nuclear fields but failed. It was at least in keeping with his own remark of 1921 that 'discovery in the grand manner is for young people and hence for me is a thing of the past.'

▶ $E=MC^2$

Fortunately, then, he had completed three other papers in his youth (in 1905) in addition to his one on the special theory of relativity! One of these included the now famous deduction which equated energy to mass in the formula $E=mc^2$ (where E=energy, m=mass and c=the speed of light). This understanding was vital in the development of nuclear energy and weapons, where only a small amount of atomic mass (when released to multiply by a factor of the speed of light squared under appropriate conditions) could unleash huge amounts of energy. The third paper described Brownian motion, and the final paper made use of Planck's quantum theory in explaining the phenomenon of the 'photoelectric' effect, helping to confirm quantum theory in the process.

FURTHER ACHIEVEMENTS

Almost inevitably, Einstein was also drawn into the atomic bomb race. He was asked by fellow scientists in 1939 to warn the US President of the danger of Germany creating an atomic bomb. Einstein himself had been a German citizen, but had renounced his citizenship in favour of Switzerland, and ultimately America,

having moved there in 1933 following the elevation of Hitler to power in his home country. Roosevelt's response to Einstein's warning was to initiate the Manhattan project to create an American bomb first.

After the war Einstein spent time trying to encourage nuclear disarmament.

ALEXANDER FLEMING

1881–1955

• **1929** Fleming publishes first report on antibacterial properties of penicillin • **1939** Indirectly provides penicillin to Howard Florey and Ernst Chain • **1944** Becomes Sir Alexander Fleming • **1945** Awarded the Nobel Prize for Medicine jointly with Florey and Chain • **1955** Dies of a heart attack in London

Alexander Fleming had had a quite unremarkable life up until the chance discovery of a mould in his laboratory in September 1928. Even the decade after he made the find which would go on to save millions of lives, little changed. It took until 1940 for the 'penicillin' to be produced from the fungi in practically useful quantities. Although it was in fact another team of people altogether who facilitated the latter development, it was Fleming who became revered as a hero.

The son of a Scottish farmer, Fleming came from humble background, and began his working life at the age of sixteen as a shipping clerk in London, England. After inheriting a small amount of money, and following suitable encouragement from this brother who was a doctor, Fleming decided to study medicine. In 1902, he joined St. Mary's Hospital Medical School in London, where he remained for the rest of his career barring a period from 1914-18 putting his medical skills to good use for the war effort.

▶ AN INTEREST IN BACTERIOLOGY

Fleming became increasingly interested in bacte-

The discovery of penicillin was due as much to luck as to scientific study

riology. Indeed, it was his wartime experiences which made him realise there was a need for a non-toxic drug to combat the millions killed by the bacteria which infected wounds. After he rejoined St Mary's therefore, he searched for a naturally occurring bacteria-killer and focused initially on what he believed were the body's own sources: tears, saliva and mucus from the nose. In 1922, he had his first success, producing lysozyme, an enzyme produced by the body. It killed certain bacteria naturally, but Fleming could not produce it in sufficiently concentrated quantities to be of medical use.

▶ A STROKE OF LUCK

The search continued, although even scientists sometimes have to take a holiday, and ironically it was a two-week break which led to Fleming's ultimately world-changing discovery. Before leaving for his vacation in 1928, however, the bacteriologist had been examining some dishes containing staphylococcus bacteria, which turned out to be the first in a sequence of rather fortunate events. He accidentally left one of the dishes exposed to the air before he departed and it became infected with Penicillium notatum. The form of infection in itself was lucky as it was only because it was being studied elsewhere in the hospital that it was present to contaminate Fleming's sample at all, and a cold spell of weather in Fleming's absence allowed the fungi which developed to grow.

Although Fleming had luck on his side in the first instance, the fact he was a skilled bacteriologist was also vital. On returning from his holiday, he noticed a mould had grown in the infected dish and, rather than simply wash it out, was sufficiently interested to examine it further. He noticed clear patches around the edges of the contamination and correctly deduced that there was something in the Penicillium notatum which was killing the staphylococcus bacteria. On further testing he found it was a useful killer of many forms of bacteria, but again it occurred in sufficient quantities to be of much further use.

▶ THE DEVELOPMENT OF PENICILLIN

So it was left to the spur brought about by World War Two over a decade later and a new team of scientists before the quest for a non-toxic antibiotic was revived and 'penicillin' as Fleming had named his finding was revisited. The Scot supplied the team led by Howard Walter Florey and which included a chemist called Ernst Boris Chain with a sample of his mould. By 1940 the team had proved penicillin's potency in fighting infections in mammals and soon afterwards made the breakthroughs necessary for it to be produced on an industrial scale.

The importance of Florey and Chain to the story was at least acknowledged when, along with Fleming, they were jointly awarded the Nobel Prize for Physiology in 1945.

FURTHER ACHIEVEMENTS

Natural and semi-synthetic versions of penicillin would go on to be mass-produced, saving millions of lives during the war and even more afterwards as it was used to combat a whole series of bacteria-causing diseases. Fleming would be hailed as a saviour by a public in need of heroes and was knighted in 1944, although the team of later scientists had arguably done most to make penicillin useful. Fleming himself said of his role, 'My only merit is that I did not neglect the observation and that I pursued the subject as a bacteriologist.'

ROBERT GODDARD

1882–1945

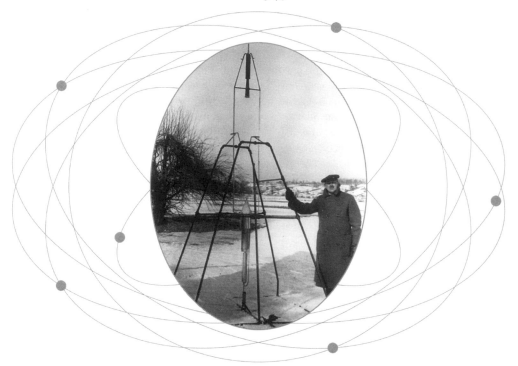

• 1915 Demonstrates rocket engines can produce thrust in a vacuum •
1926 Launches first liquid-fueled rocket, which reaches an altitude of almost 15 metres
• 1930 Starts working in Roswell, New Mexico, where he develops supersonic and multi-stage
rockets and fin-guided steering

'It has often proved true that the dream of yesterday is the hope of today and the reality of tomorrow.' (Robert Hutchings Goddard, 1904). This is probably not the measured reflection of the average high school leaver, but then the pioneer of modern rocketry was never going to be an average student. From the age of seventeen, the American Robert Goddard had known exactly what he was going to do with his life. One October day in 1899, he had had an all-consuming vision while chopping branches off a cherry

tree in his garden of '...how wonderful it would be to make some device which had even the possibility of ascending to Mars. I was a different boy when I descended the tree from when I ascended, for existence at last seemed very purposive.'

▶ OF SCEPTICS AND IGNORANTS

A few years later when the *New York Times* found out about Goddard's vision, the scientist was upset to find that the newspaper not only failed to share it but also mocked him for it. By this stage, Hutchings had already written a paper entitled *A*

'Every vision is a joke until the first man accomplishes it'

Method of Reaching Extreme Altitudes which he had published in 1919, outlining his advances in rocketry to date and his hopes of a future landing on the moon. The 13 January 1920 editorial in the *New York Times* scoffed at the doctor of physics (as he was by then), teaching at Clark University in Worcester, Massachusetts. It rebuked him for lacking 'the knowledge ladled out daily in high schools' that it would be impossible for a rocket to move forward outside of the earth's atmosphere because there was no atmosphere for it to push against in order to gain propulsion.

If the column writer had bothered to delve a little deeper he would have found that Goddard had already gone a long way towards refuting exactly this objection. As early as 1907 he had completed mathematical calculations to show a rocket could thrust in a vacuum, and had backed this up with a physical experiment showing just such a concept in 1915. Internationally, others too were already envisaging space travel and were beginning to undertake work to that end, most notably a Russian called Konstantin E. Tsiolkovsky (1857–1935) and the German Hermann Oberth (1894–1989).

▸ GODDARD'S RESPONSE

The criticism directed against Goddard by the editorial merely spurred him on further. Responding to the article with the determined statement, 'Every vision is a joke until the first man accomplishes it,' he soon afterwards started making strides towards the rocket which would lay the foundations for space travel. He began to work with liquid fuels, rather than gunpowder, realising this was the most likely method by which he could power his dreams. By 1925 he had developed a prototype rocket fuelled by gasoline and liquid oxygen which succeeded in lifting its own weight for the first time in a controlled test. Just three months later, on 16 March 1926, the world's first full launch of a liquid-fuel propelled rocket took place. At his 'aunt' Effie's farm in Auburn, Massachusetts, Goddard sent a ten foot rocket just 41ft into the air and 184ft in distance over two and a half seconds. But it had worked.

Over the next decade, and thanks to better funding from the Guggenheim family for his pioneering work after 1930, Goddard improved and successfully launched over thirty more rockets, gradually increasing their altitude and reliability. He filed patents for better control, better guidance and better fuel pump mechanisms. By 1935 he had launched a rocket which travelled faster than the speed of sound. His efforts culminated in a launch on 26 March 1937 with a rocket which reached 1.7 miles in altitude, then a record.

Yet in spite of his successes, the US government largely ignored Goddard's work until the space race gathered full momentum in the 1940s and 1950s. It turned to Goddard's developments as a base from which to begin. Indeed, the government was eventually forced to pay one million dollars for patent infringement to Goddard's widow in acknowledgement of the use they had made of his designs.

GODDARD VINDICATED

By the time man landed on the moon in 1969, Goddard had long since passed away. But the 'Correction' to the 1920 editorial by the New York Times *three days before Neil Armstrong's historic first lunar steps vindicated the man who had played a large part in ultimately putting him there. 'It is now definitely established,' the newspaper wrote, 'that a rocket can function in a vacuum as well as in an atmosphere. The Times regrets the error.'*

NIELS BOHR

1885–1962

Few twentieth century theoretical physicists are referred to in the same breath as Albert **Einstein** (1879–1955) but the Dane Niels Bohr is one of them. He made major contributions to validating the concept of quantum physics set out by Max **Planck** (1858–1947) in 1900, solved issues concerning the behaviour of electrons in Ernest **Rutherford's** atomic structure and was involved in the development of the first atomic bomb.

▶ THE COLLAPSING ATOM

Bohr gained his PhD at the University of Copenhagen in 1911 then moved briefly to the Cavendish Laboratory at Cambridge before settling in Manchester to work with Ernest Rutherford. The New Zealand physicist had just established his 'planetary' model of the atom: a tiny central nucleus bore most of the weight, around which electrons spiralled in a series of orbits. But there was a problem with this model.

Electrons only existed in 'fixed' orbits where they did not radiate energy

Classical physics insisted that if the electrons moved around the atom in this way, the energy they radiated would ultimately expire and the electrons would collapse into the nucleus. In 1913 Bohr resolved this issue and simultaneously validated Rutherford's model by applying Planck's quantum theory to it. He argued, from the perspective of quantum theory, that electrons only existed in 'fixed' orbits where they did not radiate energy. Quanta of radiation would only ever be emitted as an atom made the transition between states and absorbed or released energy. Only at this point in time would electrons 'move,' hopping from a lower to a higher-energy orbit as the atom took on energy, or jumping down an orbit as it emitted it (producing light in the process).

Bohr calculated the amount of radiation emitted during these transitions using Planck's constant. It fitted physical observations. Also, when he applied this to hydrogen atoms and the wavelengths of light that they should have released under this principle, he again found his calculations matched real world examples. It was a bizarre concept to grasp, as had been Planck's initial enunciation of quantum theory, but here was another practical example which validated it.

▶ CORRESPONDENCE AND COMPLIMENTARITY

Bohr made another important addition to quantum theory and one to the school of quantum mechanics which succeeded it. In the former in 1916 he enunciated the 'correspondence principle': in spite of the huge apparent differences between the two, the laws which govern quantum theory at the microscopic level should still correspond with our understanding of classical physics as observed on the larger 'real world' level.

Later in 1927, in quantum mechanics, Bohr added the 'complementarity principle'. This argued that debates over whether light, as well as other atomic objects, behaved in a wave-like or particle-like fashion were futile because the equipment used in experiments to try to prove the case one way or the other greatly influenced the outcome of the results. Instead all results only gave a partial glimpse of the answer to any atomic test and therefore had to be interpreted side-by-side with all other results to give a broader 'sum-of-the-parts' understanding. This idea sat neatly beside theories offered by the likes of Louis de Broglie (1892–1987), Werner **Heisenberg** (1901–76) and Max Born (1882–1970).

THE LEGACY OF BOHR

Bohr's theoretical and practical involvement in the physics which led to the creation of the first atomic bomb was dramatic. In 1939 he developed a theory of nuclear fission (splitting a heavy atom's nucleus to release huge amounts of energy suitable for an atomic bomb) with John Archibald Wheeler (b.1911) from Bohr's 'liquid-drop' description (1936) of the way protons and neutrons bonded in the nucleus. He realised, ominously, that the uranium-235 isotope would be more susceptible to fission than the more commonly used uranium-238.

Ultimately his findings would make their way to the US atomic bomb project, especially after Bohr escaped to America to flee occupied Denmark and acted as a consultant to the team.

Bohr was, however, uncomfortable with the implications of the new technology and dedicated much of the rest of his life to encouraging the control and limitation of nuclear weapons, founding the Atoms for Peace Movement for physicists with similar opinions to his own.

Bohr was awarded Nobel Prize for Physics in 1922.

ERWIN SCHRÖDINGER

1887–1961

CHRONOLOGY • **1908** Schrödinger enters University of Vienna: studies mathematics and theoretical physics under Mertens and Wirtinger • **1926** Publishes his paper in which he outlines the elements of quantum wave mechanics • **1927** Joins the faculty of the University of Berlin, along with Albert Einstein • **1944** Publishes *What is Life?*

The mid-1920s was open season in the field of quantum theory and one of the many physicists who waded in with a new influential direction was Erwin Schrödinger. Born in Vienna, to a prosperous merchant family, Schrödinger had a grandmother who was half Austrian, and half English, the English side of the family originating from Leamington Spa, and he grew up speaking both English and German in the home. Schrödinger was taught at home by a private tutor until he was ten. The Austrian scientist developed what became known as 'wave mechan-

ics,' although like others, including Einstein, he later became uncomfortable with the direction quantum theory took after doing so much in the first place to validate it.

▸ NOT PARTICLES BUT WAVES

Schrödinger's own development was built largely on the back of the 1924 proposal by Louis **de Broglie** (1892-1987) that particles could, in quantum theory, behave like waves. Whilst the Austrian Schrödinger was attracted to this explanation, he was troubled by certain implications of it. Essentially, he felt de Broglie's equations

Rejecting the idea of particles, Schrödinger argued everything was a form of wave

were too simplistic and did not offer a detailed enough analysis of the behaviour of matter, particularly at the subatomic level. So he took things a stage further and removed the idea of the particle completely! In its place, he argued everything was a form of wave.

▶ A WAVE EQUATION

Amazingly, between 1925 and 1926 he was able to calculate a 'wave equation' which mathematically underpinned this argument and the science of quantum wave mechanics was born. Further proof came when the theory was applied against known values for the hydrogen atom, and correct answers were obtained, for example, in calculating the level of energy in an electron. It clearly overcame some of the more woolly elements of the earlier quantum theory developed by Niels **Bohr** (1885-1962) and addressed the weaknesses in de Broglie's thesis.

▶ ADAPTING WAVE THEORY

Indeed, the theory behind wave mechanics was now applied to all sorts of other situations with great effect. Unfortunately, it too had some fundamental weaknesses and Schrödinger was not blind to these. The overriding one was, having done away with particles, it was difficult to offer a physical explanation for the properties and nature of matter. The Austrian came up with the concept of 'wave packets' which would give the impression of the particle as seen in classical physics, but would actually be a wave. The justifications he offered, though, were found not to add up.

▶ THE PROBABILISTIC INTERPRETATION

This left Schrödinger's work susceptible to being superseded by that of others, just as his had improved on those whose ideas came before him. Shortly afterwards, the probabilistic interpretation of quantum theory based on the ideas of **Heisenberg** (1901-76) and **Born** (1882-1970) took hold. This effectively proposed matter did not exist in any particular place at all, being everywhere at the same time, until one attempted to measure it. At that point the equations they put forward offered the best 'probability' of finding the matter in a given location. Whilst this is still widely accepted as the most adequate explanation today, Schrödinger joined **Einstein** and others in condemning such a loose, probabilistic view of physics where nothing was explainable for certain and essentially cause and effect did not exist.

Ironically, Paul Adrien Maurice Dirac(1902-84), another important influence in quantum mechanics, went on to prove Schrödinger's wave thesis and the alternative probabilistic interpretation he abhorred were mathematically, at least, the equivalent of each other. Schrödinger shared a Nobel Prize for Physics in 1933 with Dirac.

SCHRÖDINGER'S CAT

This famous 'animal' is in fact part of a thought experiment designed by Erwin Schrödinger in the 1930s, to try and explain the problem that, contrary to all logic, atoms can exist in two states simultaneously, decayed and undecayed. Schrödinger uses the analogy of a cat locked in a box with a vial containing deadly poison. The vial's lid contains a radioactive atom. If the atom decays, the released particle opens the vial and the cat dies. This is an example of a quantum system, in which it seems the cat exists in an indeterminate state, because the atom is both decayed and undecayed, implying that the unobserved cat is neither dead nor alive, which is patently absurd.

HENRY MOSELEY

1887–1915

CHRONOLOGY • 1910 Appointed Lecturer in Physics at Manchester University
• 1913 Publishes first paper on X-rays, containing the elements of
what will become known as Moseley's Law • 1914 Publishes paper asserting existence of three
elements between aluminium and gold • 1915 Moseley is killed during the Gallipoli landings in
Turkey during the First World War

There were a number of future Nobel Prize winners working with Ernest **Rutherford** at Manchester University in the few years before the outbreak of the First World War. One who could well have gone on to win such an award but was robbed of a promising scientific career by the war itself was the Englishman Henry Gwyn Jeffreys Moseley. In spite of only a few years given to him to make any kind of academic progress at all, though, he still had time to achieve enough to place his name amongst the scientific greats.

▶ A SCIENTIFIC BACKGROUND

Moseley was born into an academic family, his father a leading anthropologist and zoologist and his grandfather, as well as being a priest, noted for his efforts in physics, mathematics and astronomy. So, after growing up in Weymouth, Dorset, it was little surprise Moseley proved capable enough to go on to Oxford University to study natural science. He graduated from there in 1910 and was keen to follow in the family's scholarly tradition by immediately joining Rutherford's team in Manchester.

Killed at Gallipoli aged just 28, Moseley did enough to ensure his place among the greats

Initially, like so many of the others there he worked on trying to further understand radiation, particularly of radium. Moseley soon became interested by x-rays, however, and learning new techniques in measuring their frequencies. A method had been developed for placing metals in an x-ray tube and using crystals to diffract the emitted radiation, which had a wavelength specific to the element being experimented upon.

▶ EXAMINING X-RAY SPECTRA

In 1913, Moseley examined the X-ray spectra of more than thirty metallic elements and recorded the frequencies of the lines produced. He noted the lines moved according to the elements' atomic weight. Moreover, he soon deduced the frequencies of the radiation produced were related to the squares of certain incremental whole numbers. These integers were in themselves indicative of the 'atomic number' of the elements, and therefore their position in the periodic table. In addition, Moseley noted this atomic number was the same as the positive charge of the nucleus of an atom (and by implication also the number of electrons with corresponding negative charge in an atom).

By uniting the charge in the nucleus with the atomic number, and therefore an element's position in the periodic table, Moseley had found a vital link between the physical atomic make-up of an element and its chemical properties (as indicated by its position in the periodic table). Indeed, the step change meant the properties of an element were now considered much more in terms of its atomic number than its atomic weight, as had previously been the case.

▶ RE-EXAMING MENDELEEV

By realigning the periodic table according to this atomic number rather than the atomic weight, certain inconsistencies in the **Mendeleev** version could be ironed out in a logical fashion. Most notably, 'Moseley's law' (the principle he expressed outlining the link between the X-ray frequency of an element and its atomic number), predicted that there were several missing elements in the periodic table as he improved and restructured it according to his findings. Naturally, he was able to forecast the undiscovered elements' atomic numbers, weights and other properties from their expected position in the table. Over the ensuing years, the absentees were found and they slotted into their designated places.

A LIFE CUT SHORT

Sadly, Moseley did not live long enough to see his predictions come true. On the outbreak of the First World War he signed up with the Royal Engineers. He was killed the following year by a sniper's bullet through the head at the Battle of Suvla Bay in Gallipoli. At the age of just twenty-seven science had lost one of its brightest young sparks, already widely acknowledged for his completed work, and with the potential for further proving his genius cruelly snatched away in a single shot.

EDWIN HUBBLE

1889–1953

• 1919 Hubble joins the staff of the Mount Wilson Observatory
• 1923 Proves that the universe extends beyond the edges of our home galaxy, the Milky Way • 1925 Creates the first useful scheme for classifying galaxies
• 1929 Demonstrates that the universe is expanding • 1936 Publishes *The Realm of the Nebulae,*the most popular science book of the year • 1990 The giant orbiting Hubble telescope is named in his honour

The man who completely changed our view of the bubble in which we exist was almost lost to astronomy, first to boxing, then to law. The young Hubble was such a fine fighter during the days of his astronomy and mathematics degree at the University of Chicago boxing promoters tried to persuade him to turn professional. He refused the offer. He did not turn down the chance to go to Oxford University in the United Kingdom on a Rhodes Scholarship to study law in 1910, though.

He duly gained a BA in 1912 and contemplated a career in law on returning to the United States. In comparison to astronomy, he found the subject boring, however, so instead returned to Chicago to gain his PhD in the field of study he loved. After serving and being injured in the First World War, he finally had the chance to observe the stars professionally, taking up a post in the Mount Wilson Observatory in California in 1919, where he would spend the rest of his career.

The astronomer was lucky in that shortly after

As Hubble measured the distances of galaxies from the Earth, he found they were receding

he arrived, the observatory built a brand new 100-inch telescope, which was the most powerful in the world at that time. It allowed Hubble to view the skies in a level of previously unseen detail. He quickly took full advantage of this privilege. The American was particularly interested in the many 'nebulae' in the skies, all of which were thought to be clouds of dust within our own Milky Way galaxy. Indeed, it was thought at the time there was only this one galaxy in all, which according to the measurements of **Halley's** contemporary and rival, Harlow Shapley (1885-1972), was approximately 300,000 light years across (this was subsequently revised to 100,000 light years).

Focusing on the Andromeda nebula, Hubble used a technique developed by Shapley himself to ascertain that this 'cloud' was some 900,000 light years away from earth and therefore clearly outside the Milky Way. Moreover, Hubble soon came to realise these spiral-shaped nebulae were in fact other galaxies, much like our own. There were literally millions of them in the sky, containing billions of other stars. The results were breathtaking, completely changing our perception of the size of the universe, and brought Hubble fame overnight.

▸ THE GALAXIES RECEDE

Moreover, during the next few years, Hubble continued measuring the distances of the galaxies from earth and found they seemed to be moving away from it, or 'receding.' In addition, the greater the distance between the earth and the galaxy, the faster the latter seemed to be receding. By 1927, Hubble came to the only logical conclusion: the universe, which most astronomers had believed was static, was in fact expanding. Other scientists had for the first time hinted at this possibility a few years earlier but now Hubble had provided conclusive evidence. Indeed, **Einstein** himself had developed an earlier theory which required the universe to be moving either inwards or outwards for it to work, but had changed it because astronomers had told him the universe was definitely static. He later referred to this alteration, on hearing the universe was actually expanding, as 'the greatest blunder of my life.'

▸ HUBBLE'S CONSTANT

By 1929, Hubble had measured the distances of enough galaxies to announce his formulation of 'Hubble's constant.' He had worked out the speed at which the galaxies were recessing to be distance multiplied by his constant. Although Hubble overestimated the size of the constant, his formula was valid. Corrections since have allowed astronomers to estimate the radius of the universe to be a maximum of 18 billion light years and its age to be between 10 to 20 billion years old. Hubble went on to provide a system of classifying galaxies which is still largely in use.

HUBBLE'S LEGACY

Edwin Hubble, most widely known today because of the space telescope named after him, revolutionised our understanding of the cosmos. In the same way that that telescope hoped to improve our perception of the universe after its launch in 1990, so the American provided the most incredible 'picture' of space that humans had ever known, some sixty-five years earlier.

The notion of a universe which was expanding allowed later scientists to, amongst other things, find consensus on the origin of space and settle on the big-bang theory. Indeed, the principle of an expanding cosmos has been at the heart of astronomical theory ever since.

SIR JAMES CHADWICK

1891–1974

CHRONOLOGY • 1911–1913 Chadwick graduates from Manchester Universtity, and spends the next two years working for Ernest Rutherford • 1913 Goes to Berlin to study under the renowned Hans Geiger • 1920 Rejoins Rutherford at the Cavendish Laboratory in Cambridge • 1932 Discovers the neutrino • 1935 Chadwick awarded the Nobel Prize for Physics

The Englishman James Chadwick had a distinguished career in physics, primarily as an assistant to Ernest Rutherford (1871–1937), even before he made the breakthrough which secured his entry into this book. His solving of one of the last remaining mysteries concerning the basic structure of the atom through his discovery of the neutron, however, saw his elevation from little known researcher to celebrated physicist.

In 1910, Chadwick rejoined Rutherford, having previously worked for him in Manchester, when the latter took on his role as head of the Cavendish Laboratory at Cambridge. They worked successfully there together until 1935, when Chadwick left to become professor of physics at Liverpool University. Chadwick's earlier work at Cambridge largely involved the showering of elements with alpha particles to see the transmutation and other effects this would have. An important spin off of this was the deduction that the nucleus of the hydrogen atom, the positively charged proton with an atomic weight of one,

Chadwick's discovery of the neutron made possible the development of the atom bomb

was actually present in the nucleus of every other atom, just in larger quantities.

▶ THE WEIGHT OF AN ATOM

This work still left difficulties, though, in explaining the atomic weight of atoms. Chief amongst them was the fact the mass of known components of an atom simply did not add up. Protons seemed to account for around half of the weight and were matched in number by an equal amount of negatively charged electrons to counter their positive charge. But the weight of an electron was one-thousandth that of a proton, so still approximately half of the atomic weight of elements was unaccounted for. One leading theory suggested the missing mass was also that of protons whose charges, and therefore they themselves, were 'hidden' by additional electrons bedded inside the nucleus. The problem with this idea, though, was when nuclei fell apart, there was no evidence for these additional electrons.

▶ A SHOWER OF ALPHA PARTICLES

Finally, Chadwick solved the conundrum in 1932 after reinterpreting the results of an experiment carried out by Irène and Frédéric Juliot-Curie (Irène was Pierre and Marie **Curie's** daughter). The couple had found in 1932 that when the element beryllium was showered with alpha particles, the resultant radiation could force protons out of substances containing hydrogen.

They had concluded the emission causing this reaction was made up of gamma rays, but Chadwick soon proved there was no way the rays could do this. Rather, it was far more likely neutrally charged subatomic units, which he named neutrons, with the same weight as protons could force this reaction and therefore were what made up the radiation. Rutherford had hinted at the existence of a similar particle back in 1920 but now there was strong evidence for it. The explanation was widely accepted and at last the riddle of atomic weight had been solved: a similar number of neutrons to protons in the atom of an element would make up the remaining fifty per cent of the previously 'missing' mass.

A knighthood followed for Chadwick in 1945, partly for this discovery but largely for his scientific service to Britain during the Second World War. Indeed, Chadwick's career was greatly affected by both World Wars but in vastly different ways. He had been robbed of four years of his scientific development, spending the First World War imprisoned in a racing stable after having had the misfortune of being in Germany to undertake work with Hans Geiger (1882–1945) at the outbreak of hostilities. The next time round, however, he spent it mostly in the USA as an effective head of the British delegation working on the development of the atomic bomb.

Chadwick received the Nobel Prize for physics in 1935 for his discovery of the neutron.

FURTHER ACHIEVEMENTS

Chadwick's discovery of neutrons – elementary particles devoid of any electrical charge – was crucial to the development of nuclear physics. In contrast with the helium nuclei (alpha rays) which are charged, and therefore repelled by the considerable electrical forces present in the nuclei of heavy atoms, the neutron is capable of penetrating and splitting the nuclei of even the heaviest elements, creating the possibility of the fission of uranium-235. This made possible the atomic bomb. For this epoch-making discovery Chadwick was awarded the Nobel Prize for Physics in 1935.

FREDERICK BANTING

1891–1941

CHRONOLOGY • **1916** Banting graduates MD from Victoria College, Toronto, and joins the Canadian Army Medical Corps • **1918** Awarded the Military Cross for gallantry in action and invalided out of the army • **1921** With Charles Best, begins study of the role of the pancreas in diabetes • **1923** The pair produce and patent insulin; pharmaceutical firm Eli Lilley begin industrial production of insulin; they are awarded the Nobel Prize for Physiology • **1939** Joins Canadian army medical unit at outbreak of Second World War • **1941** Killed when his plane crashes en route to Britain from Newfoundland

Up until the 1920s diabetes had been a ruthless killer. In 1921, after only a few months experimentation, the Canadian Frederick Grant Banting led a breakthrough in its treatment and virtually overnight offered the possibility of saving millions of lives.

Banting had graduated in medicine in 1916 from Victoria College in Toronto. In 1920, after returning from the First World War decorated with the Military Cross for bravery, Banting founded a practice in London, Ontario. At the same time he undertook research work at the local medical school, focusing on studies related to the pancreas.

▶ INITIAL RESEARCH

Earlier research had shown there was almost certainly some link between the pancreas and diabetes, but at the time it was not understood

In the age of AIDS and Ebola, one can forget that diabetes too was once a killer disease

what this was. We now know a hormone within the pancreas controls the flow of sugar into the blood stream. Diabetics are lacking this function and, untreated, are gradually killed by uncontrolled glucose input into the body's systems. Banting did not have the confirmation of this knowledge but he suspected the cause might be something to this effect.

His specialisation within pancreatic studies was in areas called the islets of Langerhans. Banting believed that the islets might be the most likely to produce the kind of hormone, if it existed, which controlled the glucose levels in the body. He figured that if this hormone could be extracted it might be viable as an injected treatment to diabetes sufferers.

▶ BANTING AND BEST

In 1921, Banting began a series of tests along with Charles Herbert Best (1899–1978) who was a research assistant at the University of Toronto, after being put in touch with him by Professor John James Rickard MacLeod (1876–1936) who also worked at the university. The two investigators were assigned a laboratory at the university by MacLeod and some dogs upon which to experiment. They extracted matter from the islets of Langerhans in the dogs' pancreas after preventing other pancreatic fluids from entering, in an effort to extract as pure samples as possible. The scientists then removed the pancreas from some dogs to hopefully induce diabetes. This soon happened,

so their next step was to try to treat the dogs with their extract. It worked. The symptoms of the disease were soon under control.

▶ PRODUCTION OF INSULIN

Following this success, Banting and Best, at the suggestion of MacLeod, decided to further purify their extracted treatment before testing it on humans. This task was assigned to James Bertram Collip (1892–1965), a biochemist, and the solution he produced was named insulin. Human trials took place in 1923 with immediate impact. Dying patients were restored to health and suddenly diabetes could be managed within the realms of a normal lifestyle. The same year, industrial production of insulin began from pigs' pancreas and patients around the world soon received the life-saving benefits.

Banting was awarded the 1923 Nobel Prize for Physiology. It was shared, rather unfairly, with MacLeod, who had only contributed a limited amount to the discovery, but not with Best, who had been actively involved. To redress the balance a little, Banting shared his portion of the prize money with Best and MacLeod with Collip. The heroics Banting displayed during the First World War were called upon again in the Second as he undertook dangerous research into the effects of poisonous gases. Sadly, this time he succumbed, not to the gases, but in an air crash while flying from Canada to the United Kingdom to share his research with the British.

A SINGLE DISCOVERY SAVES MILLIONS OF LIVES

While many of the scientists in this book are known for a number of discoveries or inventions, others are equally celebrated for just one. This is the case with Sir Frederick Banting. When one thinks of killer diseases, one tends to think of AIDS, or Ebola. It is difficult to think

of diabetes in the same category: and yet, prior to the discovery of insulin, diabetes routinely meant a death sentence to the millions unlucky enough to have it. Thanks to Sir Frederick Banting, this is no longer the case.

LOUIS DE BROGLIE

1892–1987

CHRONOLOGY • **1913** De Broglie graduates • **1914** Conscripted into the French army, where he remains until the end of the war in 1918 • **1924** At the Faculty of Sciences at Paris University delivers a thesis *Recherches sur la Théorie des Quanta* (Researches on the quantum theory) • **1927** Demonstrates the wavelike properties of electrons and other subatomic particles • **1929** Awarded Nobel Prize for Physics for his work on subatomic particles • **1952** Awarded the first Kalinga Prize by UNESCO for his efforts to explain modern physics to the layman • **1987** Dies in Paris, aged 95

Louis de Broglie probably had one of the more unusual family histories in quantum physics, with a name to match. The 'Prince' element of his title reflected his an honour bestowed upon his ancestors for service to the Austrians during an earlier war. But de Broglie was primarily a Frenchman of aristocratic background, which also meant he later inherited the title of 'Duc' upon becoming the head of his family on the death of his father.

▶ FROM HISTORY TO SCIENCE

His initial studies in history had been followed by a period working at the radio station based at the Eiffel Tower during the First World War, meaning it was a circuitous route to physical acclaim. His job at the famous landmark was the stimulus for his interest in science, and led to de Broglie signing up for study in physics at the Sorbonne after the war. This was to have a major impact.

If waves could behave like particles, why could not particles behave like waves?

▸ OF WAVES AND PARTICLES

De Broglie may have taken some time to turn to science, but he made a swift impression. Indeed, it was his 1924 doctoral thesis which formed the basis of his fame. The key theme concerned a natural extension of the quantum theory which had already been put in place. **Einstein** (1879-1955) had suggested in one of his 1905 papers the mysterious 'photoelectric' effect could be explained by an interpretation which included electromagnetic waves behaving like particles; indeed the waves were in fact constructed from a stream of particles (called 'quanta' or 'photons'). What de Broglie did was simply turn this on its head: if waves could behave like particles, why should particles not behave like waves?

▸ ELECTRON BEHAVIOUR

His conclusion was indeed they could and he formulated a theoretical proof of his idea involving the behaviour of electrons. In classical physics, these were unquestionably considered to be particles, distinct pieces of physical matter. By applying quantum theory, however, de Broglie was able to show an electron could also act as if it were a wave with its wavelength calculated by simply dividing 'Planck's constant' by the electron's momentum at any given instant. Although the proposal sounded extremely theoretical it was, remarkably, found to be plausible by experimental evidence shortly afterwards.

▸ WAVE–PARTICLE DUALITY

As a result it sparked a fierce debate concerning the 'wave-particle duality' of matter and questions about which interpretation was correct. **Schrödinger** (1887-1961), **Heisenberg** (1901-76) and **Born** (1882-1970), amongst others, would soon offer compelling arguments. To an extent, Niels **Bohr** (1885-1962) provided some context around the issue in 1927 by stressing the futility of the debate, pointing out the equipment used in experiments to try to prove the case one way or the other greatly influenced the outcome of the results. A principle of 'complementarity' therefore had to be applied which suggested all of the experimental proof one way or the other to be a series of partially correct answers which had to be interpreted side-by-side for the most compete picture. Eventually, though, the 'probabilistic' theories of Heisenberg and Born largely won out. At this juncture, the point at which cause and effect had logically been removed from atomic physics, De Broglie, like Einstein and Schrödinger, began to question the direction quantum theory was taking, and rejected many of its findings.

FURTHER ACHIEVEMENTS

There were countless physicists wading into the debate surrounding quantum theory in the mid-1920s. Most of the key contributors were established professors, doctors or specialists in physics which makes Louis de Broglie's background as a historian all the more remarkable.

His unusual route to scientific prominence did not make him any less effective when he finally got there, however, proposing a quantum theory which would once again open a whole new branch of investigation.

ENRICO FERMI

1901–1954

CHRONOLOGY • 1923 Fermi studies under Max Born at Göttingen, Germany • 1934 Discovers slow neutrons • 1938 Awarded Nobel Prize for Physics • 1939 Escapes Europe and moves to the US • 1942 Achieves man-made nuclear chain reaction • 1949 Argues against development of the H-bomb

Enrico Fermi, arguably Italy's most talented scientist of the twentieth century and quite possibly since **Galileo,** could have had no idea of the eventual outcome of the experimental work he undertook in Rome in the mid-1930s. He was systematically working his way through the elements to study the effects on them of a neutron-bombarding technique he had discovered. Most yielded predictable, or certainly not extraordinary, results. When he arrived at uranium, the heaviest naturally occurring element, however, something very odd happened, which was to have

enormous impact on physics and beyond.

A few years later, in Chicago, Fermi would experience first hand the potential of his discovery. Fermi and his Jewish wife had fled to America following the rise of anti-Semitism in Italy.

▸ THE URANIUM NUCLEUS

Shortly afterwards, he had received reports of a reinterpretation of his uranium bombardment experiment. Fermi himself had been unsure of what had happened, suspecting the possibility that perhaps the uranium had transmuted into new, heavier elements. Now, however, an alternative

'The Italian navigator,' said one commentator, 'has landed in the new world'

explanation was offered by the German scientists Otto Hahn, Fritz Strassmann and Lise Meitner that the uranium nucleus had in fact been broken down into a number of smaller elements. Moreover, this nuclear fission had seen some of the uranium mass converted into potentially huge amounts of energy under the rules of **Einstein's** formula E=mc². This reinterpretation was leaked out of Germany by Meitner and her nephew Otto Frisch when they escaped the Nazi state.

▸ A NEW WORLD

Fermi immediately saw the impact of the analysis and set to work on reproducing the experiment with Niels **Bohr** on arrival in the US. They confirmed their best and worst fears: using the uranium isotope-235 a nuclear chain reaction could almost certainly be created as the basis of an atomic bomb. Fermi was recruited to the Manhattan Project to ensure the US created a fission bomb ahead of the Germans. Fermi led a team in Chicago seeking to generate a self-sustaining, contained nuclear reaction. By 2 December 1942 his team had created an 'atomic pile' of graphite blocks, drilled with uranium which went on to produce a self-sustaining chain reaction for nearly half an hour. 'The Italian navigator,' as one commentator reported back to the project committee, 'has just landed in the new world.' Less than three years later, the technology would be used in the first atomic bombs

with devastating effect.

The innocent discovery back in Italy in the 1930s, which had led to such incredible consequences, had been Fermi's conception of neutron bombardment in the artificial transmutation of elements. The Juliot-Curies had announced in 1934 their discovery that radioactive isotopes could be generated artificially by showering certain elements with alpha particles.

▸ ON NEUTRONS

Fermi had quickly realised that the newly-discovered neutrons would be even more suited to this purpose because their neutral charge would be more likely to allow them to slip into elements' nuclei without resistance. By chance he also found the phenomenon of 'slow neutrons' by placing a piece of solid paraffin in front of his target element during bombardment. This had the effect of slowing the neutrons down before they reached the element, meaning they were exposed to its nuclei for longer and thereby had a much greater chance of being drawn in to create new isotopes. As Fermi now worked through the elements applying these discoveries, he created lots of new radioactive isotopes, which was considered achievement enough for him to be awarded the 1938 Nobel Prize for physics. It was only after he had collected his award that the much more significant consequences of this work when applied to uranium were realised.

FURTHER ACHIEVEMENTS

Earlier in his career, Fermi had established his reputation with important work in theoretical physics. His most notable achievement in this area was his concept of radioactive beta decay. This concerned the theory that a proton could be created from a neutron via the shedding of an electron (a beta particle) and something

known as an antineutrino.

It was for his achievements in experimental physics for which the Italian would be remembered, however, leaving behind a world, following his early death from cancer, very different from the one he had entered just over half a century earlier.

WERNER HEISENBERG

1901–1954

• 1922 Heisenberg studies at Göttingen, Germany under Born • 1925 Heisenberg develops his radical approach to quantum theory • 1927 Formulates his famous 'Uncertainty Principle' • 1932 Awarded the Nobel Prize for Physics

Heisenberg's development of matrix mechanics in 1925 sparked a controversy in the rarified world of quantum theory. Like many other physicists, Heisenberg too had been contemplating the debate over whether electrons and other atomic phenomena behaved in a wave or particle-like fashion. Heisenberg found a simple solution. He ignored both arguments altogether! Instead Heisenberg proposed that the only important factor was being able to mathematically predict the occurrence of atomic features which could be measured or observed such as frequency and light emissions. So, he

applied algebra to the problem and developed a mathematically based solution which came to be known as matrix mechanics. The predictive and quantifiable powers of this new scheme were excellent and Heisenberg received the Nobel Prize for this development in 1932.

▸ THE UNCERTAINTY PRINCIPLE

They also had a logical extension and it was to this part **Einstein** objected most. Heisenberg expressed it as his 'uncertainty' principle in 1927. In seeking to underline his basis for ignoring the visual idea of the atom and only considering it mathematically, Heisenberg

Either a particle's position or its momentum can be known: not both together

realised in physical reality it was not possible to measure both the exact position and exact momentum of a particle at the same time. The reasoning behind this was simple: if one undertook an experiment to determine the position of, say, an electron at any instant, something like gamma rays would have to be deflected off the particle to locate it. In doing this, the position might be identified, but the electron's momentum would be radically changed by its interaction with the gamma rays. By the same token if less intrusive techniques were used to find the electron, the original momentum might be better preserved, but the accuracy of the positioning would be woolly. This meant the best that could be hoped for was a mathematical prediction of the probability of the electron's position and location at any given instant, and Heisenberg supplied this formula.

▶ UPSETTING CAUSE AND EFFECT

Unfortunately, the logical conclusion from accepting this is that cause and effect as relied upon in classical physics can no longer be produced. The best that can be hoped for is a series of probabilities about the behaviour of any given particle at any point now or in the future. Max Born's 'probabilistic' interpretation, expressed at about the same time and concerning the likelihood of finding a particle at any particular point through probability defined by the amplitude of its associated wave, led to similar conclusions. On hearing of the radical ideas, Einstein remarked, 'God does not play dice. He may be subtle, but he is not malicious.' Still, the approach is now largely accepted.

▶ DEVELOPING THE BOMB

Heisenberg partook in other important work, too. After **Chadwick** (1891–1974) had discovered the neutron in 1932, it was Heisenberg who suggested the model of the proton and neutron being held together in the nucleus of the atom. Moreover, Heisenberg played a significant and controversial role in Germany's attempts to develop an atomic bomb during the Second World War. Unlike many of his countrymen, Heisenberg did not leave his homeland when Hitler came to power, but by the same token he was no Nazi sympathiser. The government was well aware of Heisenberg's leading atomic knowledge, however, and compelled him to head up a team which would seek to create an atomic bomb. With the main Nazi focus being largely on developing other types of weapon, however, the team did not deliver in time to alter the course of the war. Furthermore, Heisenberg maintained after the war he had no intention of ever letting the project succeed and be in a position to hand over such a powerful device to Hitler anyway. He insisted, had it been necessary, he would have used his position to hijack the team's progress if it had come close to creating such a device.

THE INFLUENCE OF HEISENBERG

Of all the competing models of quantum theory created in the 1920s, the theories developed by Werner Heisenberg, along with proposals based on similar principles by his German countryman Max Born (1882–1970), have endured the longest. The approach which turned physical science into merely a series of unpredictable probabilities appalled one of the greatest scientists of the century, Albert Einstein, amongst others, but Heisenberg's ideas worked and as a result they continued to be accepted.

LINUS CARL PAULING

1901–1994

CHRONOLOGY • **1925** Pauling awarded PhD in Chemistry from the California Institute of Technology • **1934** Begins to study the complex molecules of living tissues, particularly in connection with proteins • **1940s** Pauling becomes interested in sickle-cell anemia • **1954** Awarded the Nobel Prize for Chemistry • **1961** Explains the molecular basis of anesthesia • **1962** Awarded the Nobel Prize for Peace

Linus Carl Pauling is particularly noted for his contributions to structural chemistry and his application of quantum theory in this area, as well as effectively founding molecular biology. In later life he achieved what only very few scientists ever succeed in, wide fame amongst the general public, principally for his anti-nuclear stance and advocacy of the health-giving properties of large quantities of vitamin C.

Pauling received his first Nobel Prize, for chemistry, in 1954, for the significant progress he had made in understanding chemical, in particular molecular, bonds. Earlier in the century, Pauling's American countryman Gilbert Lewis (1875–1946) had offered many of the basic explanations for the structural bonding between elements which are now familiar in chemistry. These included the sharing of a pair of electrons between atoms and the tendency of elements to combine with others in order to 'fill' their electron 'shells,' according to rigidly defined orbits (with two electrons in the closest orbit to the nucleus, eight in the second orbit, eight in the

'I wanted to understand the world!'

LINUS CARL PAULING, WHEN ASKED WHY HE BECAME A SCIENTIST

third and so on). Pauling now built on these efforts with research into more complicated bonds between molecules. Although he spent much of his career at the California Institute of Technology, the two years in Europe working alongside some of the finest brains in physical quantum theory in 1926 and 1927 greatly influenced his later work in structural chemistry.

▶ A NEW APPROACH TO CHEMISTRY

He was one of the first to realise the impact of this new physical interpretation to understanding the bonds and nature of molecules and crystals from a chemical perspective. It was his application of quantum theory to structural chemistry, indeed, some say effectively founding structural chemistry in the modern sense, which enabled Pauling to make great strides. He went on to gather vast amounts of quantifiable data concerning the measurements and properties of molecules and crystals. This helped establish the subject and could be applied in making further predictions, including the formulation in 1929 of a series of influential rules about the stability of molecular structures. He summarised all his ideas in this area in his 1939 book *The Nature of the Chemical Bond and the Structure of Molecules and Crystals*, which became a leading authority.

Pauling later moved into biochemistry again, effectively founding a new branch known as molecular biology through his discovery of the first 'molecular disease', sickle-cell anaemia. He also formulated theories on the immune system, provided chemical explanations for the way in which anaesthetics worked, and offered insights into the structure of proteins. He was also involved in the 'DNA race' to understand the nucleic acid's structure. Although his answer was ultimately wrong, Pauling provided context against which **Watson** (b. 1928), Crick (b. 1916) and Wilkins (b. 1916) could compare their studies as well as take advantage of some of his methodologies.

▶ A PACIFIST CONSCIENCE

Ironically, Pauling entered the public consciousness less for his chemical achievements and more for his stance against nuclear arms, and war in general. He refused to take part in the Manhattan Project during the Second World War and, indeed, his increasingly anti-nuclear stance after the war led to accusations of him being unpatriotic and to receiving some harassment from the authorities. For his efforts, though, he was awarded the Nobel Prize for Peace in 1962. Controversy continued to follow Pauling when he encouraged members of the public to take huge quantities of Vitamin C for of its alleged health-giving properties. There was limited scientific evidence for this, but the turnabout it had brought in his own health was remarkable and provided the source of his advocacy.

FURTHER ACHIEVEMENTS

Regarded by many as the most influential chemist of the twentieth century, Pauling is different from many of the scientists in this book in that he is not remembered for any single, specific world-changing theory, but more for a diverse range of improvements in biochemical understanding. With an incredible memory,

an instinctive feel for possible solutions to problems then confirmed by experimentation, and a talent for filling in the gaps between branches of science left behind by those working in specialised areas, he was the first person to receive two separate, individual Nobel Prizes.

ROBERT OPPENHEIMER

1904–1967

CHRONOLOGY • **1925–27** Oppenheimer studies at Cambridge under Rutherford and at Göttingen University with Niels Bohr and Max Born, where he receives his doctorate • **1942** Becomes director of the Manhattan Project, the US/British joint attempt to create an atomic bomb • **1945** Resigns his post after the use of the atomic bomb at Hiroshima and Nagasaki • **1953** After an unfavourable security hearing his contract at the Atomic Energy Commission is cancelled; Oppenheimer remains at odds with the system • **1963** Presented with the Enrico Fermi Award

'I am become Death, the destroyer of worlds.' Robert Oppenheimer's words, quoted from the *Bhagavad-Gita*, on seeing the first test of the atomic bomb reflected the heavy burden placed upon many of the world's leading scientists during the Second World War. The date was 16 July 1945 and the moment was the culmination of nearly three years' effort by the team Oppenheimer had led. On the one hand they had been working at the forefront of their field, trying to achieve something which had never been seen before against a pressing deadline. On the other hand, the outcome of their work was a tool of destruction on an incomprehensible level, with moral, political and military implications which many of the team members would subsequently struggle with. If there had ever been an age of innocence in science, it was obliterated at 5.30am on that morning in the New Mexico desert.

'I am become Death, the destroyer of worlds'

ROBERT OPPENHEIMER AT THE FIRST ATOMIC BOMB TESTS

▶ THE MANHATTAN PROJECT

Oppenheimer was a natural choice to head up the scientists on what became known as the Manhattan Project. From 1929–42 he had taught physics at the University of California, focusing in particular on the new developments in quantum and atomic theory. During this period he had played an important part in the discovery of the positron, a particle the same weight as the electron, but positively charged. In normal times this role alone would have been noteworthy. But Oppenheimer did not live in normal times and this contribution is now virtually forgotten, such was the dominance of later events.

▶ GATHERING TALENT

The stimulus for the gathering of the Manhattan Project team had been a letter by Albert **Einstein** (1879–1955), at the prompting of other concerned scientists, to President Franklin D. Roosevelt. This outlined the risk to mankind if the Nazis were successful in creating an atomic bomb first. The response of the government was to instruct the army to investigate ways of ensuring the U.S. and her allies made use of nuclear technology ahead of the German dictatorship. Consequently, Oppenheimer was appointed to lead the scientific team, choosing by 1943 the location of Los Alamos, New Mexico, in which to carry out the work. Many of the world's best physicists had fled to the United States following the tyranny in

Europe and Oppenheimer took advantage of this by bringing them together alongside homegrown talent, performing the delicate role of managing, motivating and moulding a successful team, while at the same time shielding them from the pressure and demands of their military superiors.

After the war, Oppenheimer continued to work with the military, serving on the General Advisory Committee of the Atomic Energy Commission as chairman. Like many who had seen the devastating effect of their work over Japan at the end of the Second World War, however, Oppenheimer had his reservations about the wisdom of the continued pursuit of the new technology. The government, though, now in a new 'Cold War' with the Soviet Union, was determined to press on with the development of an even more powerful weapon: the hydrogen bomb. The Commission opposed this move, with sad consequences for Oppenheimer. Already accused by the military of being friends with Soviet sympathisers and Communists in the 1940s, Oppenheimer was charged with disloyalty and an enquiry investigated these claims. He was never found guilty of any such accusations but still, when the hearing reported in 1954, it recommended Oppenheimer be stripped of his security clearance. It caused an outcry at the time and has emphasised starkly ever since the inescapably fine line between scientific progress, morality and politics, which groundbreaking scientists are often forced to tread.

FURTHER ACHIEVEMENTS

Oppenheimer's security clearance was never reinstated, although the rift was to some extent redressed when he received the Enrico Fermi Award in 1963. Oppenheimer of all people was in a position to know the intimate details of atomic weapons. This fact, coupled with his principled opposition to the use of atomic

weapons, made him a rallying point for the anti-nuclear lobby. Oppenheimer, and for rather different reasons Edward Teller, show that although science's methods may be impartial, the direction in which scientific research proceeds, can be shaped by the scientist's personal moral and political beliefs.

SIR FRANK WHITTLE

1907–1996

• **1931–32** Whittle flies planes as a test pilot for the Royal Air Force • **1934–37** Studies engineering at Cambridge University • **1936** With colleagues, sets up Power Jets Ltd • **1941** After it is discovered that the Germans have a jet engine, the British government performs the first trials of a British jet • **1944** The Gloster Meteor enters service with the RAF • **1948** Whittle is knighted

The Wright brothers had given the world the aeroplane in 1903 and in the process had made the world a smaller place. Despite developments of propeller aircraft over the following decades, though, air travel was still a long way from resembling the global industry which now facilitates lifestyles and vacations never even conceived of a hundred years ago. It would take a completely different type of engine, the jet engine, to change all that.

▶ A VISION OF AERONAUTICS

Frank Whittle, the Englishman who would go on to be the first to patent the breakthrough in air travel, was a natural candidate for making the development. Joining the RAF at just sixteen as an apprentice, he was exposed to the highs and lows of flying from a young age. He quickly qualified as a fighter pilot, spent time as a flight instructor, undertook assignments as a test pilot and studied mechanical sciences at Cambridge University from 1934-37, so his aeronautical experience was vast.

Whittle realised the need for an aircraft which could fly at great speed and altitude

▶ THE NEED FOR SPEED

Even before he had undertaken much of his flying training and service, however, Whittle had realised the need for an aircraft which could fly at great speed and at great altitude, in order to take advantage of the thinner atmosphere and therefore lower resistance to achieve this velocity. Propellor aircraft were not suited to this altitude, so he began to work on ways of overcoming the problem. The answer was the turbo-jet engine. He had hinted at such a concept in a thesis of 1928 and as early as 1930, Whittle had patented his first design for such a device. It involved an ingenious plan for an engine which took in air and after compression ignited it with fuel in a suitable chamber. The engine would then be boosted forward by the burning gas, which during the emission process would also rotate a rod. This was connected to a turbine which turned to draw in more air and start the whole cycle over again.

Further development of Whittle's design was slow due to limited funding and lack of government interest. Whittle was not deterred, however, and joined together with friends and associates to set up Power Jets Limited in 1936 with the aim of producing viable engines. By 1937 the first engine, the W1, was complete for testing. With war potentially imminent, the government had by now become interested in Whittle's work and sponsored much of its future development. The engine was attached to an aircraft which had been specially built for the purpose of carrying it, the Gloster E28/39, and on 15 May 1941 it undertook its maiden flight. Immediately it proved its potential, achieving top speeds of 370 miles per hour, faster than the quickest propeller driven aircraft, as well as operating successfully at altitudes of up to 25,000 feet. Development was now rapid because of the urgency caused by the Second World War but, nonetheless, it still took until 1944 before the jet aircraft were in active service for the RAF.

▶ PARALLEL DEVELOPMENT

Following the war, another inventor was also given credit with independently creating a jet-propelled aircraft. Although the German Hans Joachim Pabst von Ohain (1911–98) did not patent his first gas-turbine jet engine until 1935, several years after Whittle, he received financial backing sooner and as a result had his first test jet-powered aircraft in the skies ahead of the British inventor in 1939. Military production models were not brought into service on the German side until virtually the end of the war, though, and so made little impact. Von Ohain moved to America in 1947 and took up a position designing aircraft for the US Air Force.

THE LEGACY OF WHITTLE

The jet engine, then, had an important military application, but it was not until after the war that its impact on the world became apparent. The development and use of the engines in the business and tourist flight industry have turned journeys which a century ago took a few months into journeys of a few hours.

Whittle also helped play his part in this process after the war, acting as a consultant for air firms including the British Overseas Airways Corporation and Bristol Siddeley Engines. He also took on a post at the United States Naval Academy in Annapolis in 1977 as a research professor. Whittle was knighted in 1948.

EDWARD TELLER

B. 1908

CHRONOLOGY • **1930** PhD in Physical Chemistry from the University of Leipzig • **1931** Studies under Niels Bohr at Copenhagen • **1935** Emigrates to the USA • **1939** Joins Fermi's nuclear research team at the University of Chicago • **1943** Joins Manhattan Project under Robert Oppenheimer • **1952** Explodes the world's first hydrogen bomb • **1982–83** Advisor to the Reagan government on the Strategic Defense Initiative (Star Wars)

Very little could have affected the psyche of the world more than the atomic bombings of Japan in 1945. The size and sickening consequences of the two explosions over Hiroshima and Nagasaki were enough to convince most people that mankind's ability to destroy itself now existed. So when the hydrogen 'H-' Bomb was demonstrated by the United States in 1952, the psychological impact of the new weapon was not as great as it otherwise could have been. The physical effect, however,

was enormous; here was a bomb with the potential to be ten, a hundred, even a thousand times more powerful than the atomic version.

▸ TELLER AND THE BOMB

One of the key proponents of this new, devastating technology was a Hungarian born American called Edward Teller. Even before the atomic bomb project had been completed, Teller was advocating the development of the 'Super, ' a hydrogen fusion bomb, as opposed to the atomic nuclear bomb. The latter worked by essentially

Teller was a key advocate of the hydrogen bomb and of the 1980s 'Star Wars' project

splitting the nucleus of the heavy, uranium atom; the former as an offshoot of forcing the conversion of hydrogen to helium.

It was in fact, Enrico **Fermi** (1901–54) who first pointed out the possibility of a hydrogen bomb to Teller in 1941. The Italian suggested an atomic bomb could cause enough heat and pressure to force a 'thermonuclear' reaction of a hydrogen isotope, unleashing an even greater force. Scientific theory had hinted at this possibility for some time ever since it had been realised a decade earlier that a helium atom was slightly lighter than it should have been given its component parts. Clearly something was being 'lost' on fusion. An application of **Einstein's** $E=mc^2$ equation soon explained what; the missing mass was being converted into huge amounts of energy. It was exactly the basis upon which the sun worked, fusing hydrogen atoms into helium under great pressure and temperature and giving off the difference as radiation. Now on earth, the advent of nuclear technology offered the possibility of creating the necessary conditions under which to duplicate that process. Teller immediately became sold on the idea of the Hydrogen bomb.

▸ BIGGER AND BETTER BOMBS

Although the Hungarian-American physicist continued to work on the original atomic bomb project with his scientific colleagues for the rest of the war, he was already focusing on what he saw as the next essential step. Indeed, such was his vocal advocacy of the 'Super' that other colleagues could sometimes become frustrated at the attention he diverted away from the more pressing atomic bomb work.

After the war had finished, Teller himself was frustrated to find the authorities were not sufficiently motivated to begin work on the H-Bomb. By the end of the 1940s, however, it was becoming clear the Soviet Union was in the process of developing atomic technology and the government was now keen to maintain the US's advantage. So, in 1950, the H-Bomb project was begun in earnest, with Teller in a key role and fully primed to become its 'father,' as he was later nicknamed. Indeed, Teller had already been working on designs, although initially the team was thwarted when they turned out not to be viable. A rethink and collaborative work between Teller and a mathematician called Stanislaw Marcin Ulam (1909–86), who was also on the team, resulted in the overcoming of earlier technical difficulties. Teller later claimed the successful redesign was his while others have credited Ulam. Either way, the results of the group's efforts resulted in a thermonuclear device being ready by later in 1951, with a powerful public testing announcing its arrival to the world in 1952. Only a couple of years later, the US did indeed possess a bomb a thousand times more powerful than those dropped on Japan.

THE INFLUENCE OF TELLER

Around the same time, the leader of the first atomic bomb project, Robert **Oppenheimer** *(1904–67), was being investigated for alleged 'disloyalty' to his country. Teller angered colleagues by testifying against Oppenheimer, saying that he would feel more secure if public matters could rest in other hands. This 'betrayal,'* *on top of earlier tension during the Manhattan Project and friction over claims of credit for the H-Bomb, led to a split between Teller and many of his old colleagues. Nonetheless, his opinion remained influential in public life, most notably persuading the government to pursue the 'Star Wars' missile defence project in the 1980s.*

WILLIAM SHOCKLEY

1910–1989

Ever since the advent of the broadcasting industry, methods had been sought to increase the strength of electrical signals within receiving devices. The best that had been managed until nearly the mid-point of the century was the vacuum tube. It did the job but was unreliable, expensive to manufacture and being made of glass, was easy to damage. Its size also limited the amount to which receivers such as televisions and radios could be reduced, leaving them bulky and unwieldy. The prize to the person or company

who could improve on this state of affairs, then, was potentially huge. After the Second World War, Bell Telephone Laboratories began chasing this pot of gold in earnest.

▶ INVESTIGATING CRYSTALS

One of the key scientists involved in the project was William Shockley. Born in London, England in 1913, Shockley was the son of two American mining engineers William and Mary. After studying at both the California and Massachussetts Institutes of Technology, gaining his PhD in

The inventor of the transistor became a sadly controversial figure in his later career

1936. In the same year he joined the Bell Laboratories and was attached to the team headed by Dr CJ Davisson. Also members of this team were two of his fellow countrymen, John Bardeen (1908–91) and Walter Houser Brattain (1902–87). The team had been investigating the properties of electricity-conducting crystals, focusing in particular on the element germanium. It was in December 1947 when Bardeen and Brattain were conducting an experiment, at a time when Shockley was not in fact present, that they first were first successful in harnessing the power amplification properties of the crystal. Under Shockley's direction, however, they would go on to develop this point-contact 'transistor' into a device which was smaller, more efficient and more reliable than the vacuum tube, virtually invalidating the latter overnight. The three were awarded the Nobel Prize for Physics for this invention in 1956.

▶ SILICON VALLEY

Furthermore, by 1948, Shockley on his own had worked out the quantum theory behind the behaviour of the semiconductors and used his increased knowledge to apply it to an even more efficient design. This development would become known as a junction transistor and soon set a new standard. Immediately, the miniaturisation of technology began, culminating in transistors use on a microscopic level today within computer chips and all areas of modern electronic infrastructure.

Indeed, not satisfied with his initial developments, Shockley decided to set up in business on his own in 1955 in a bid to develop silicon transistors en masse, rather than the more frequently used germanium ones. Silicon was a much more commonly occurring element than germanium, so potentially cheaper, and could be used at much higher temperatures. Unfortunately, because of the latter fact it was much harder to melt for purification purposes than germanium which was why the rarer element had been preferred. Again, the company which could make silicon transistors cheaply would claim access to another pot of gold. As it turned out, Shockley's company was not successful in this, but some of his former employees who later set up on their own were. Moreover, his choice of location in San Francisco heralded the beginning of the now world famous Silicon Valley.

▶ THE RACE CONTROVERSY

After 1965 Shockley's public reputation declined following his adoption of a controversial stance on race. He left the electrical industry and began working on hereditary theories of intelligence. He concluded Caucasians were innately of higher intelligence than other races, and publicly advocated that those of lower IQs should be paid to undergo sterilisation to prevent them 'diluting' the intellectual evolution of the human race. Consequently, he is often remembered much more for his exposition of these views than his earlier scientific breakthroughs.

A FIGURE OF CONTROVERSY

William Shockley should be remembered simply for his role in the invention of the transistor in 1947, and his subsequent improvements to it. Sadly, his name is associated with controversial views concerning race as much as it is for his role in helping to define modern electronics, and indeed the modern world, as we now know it.

ALAN TURING

1912–1954

CHRONOLOGY
- **1931** Turing graduates from King's College, Cambridge
- **1937** Describes the 'Turing machine', a hypothetical computer
- **1939** Returns to Britain from Princeton University to work on code-breaking at Bletchley Park
- **1940** Creates the Bombe, a machine capable of deciphering the German Enigma code

The visionary of the modern era, Alan Turing, a scientist whose name will remain inextricably linked with the computer, was in fact first and foremost an outstanding mathematician. Indeed, it was in response to the mathematical 'incompleteness theorem' proposed by Kurt Gödel (1906–78), that Turing devised a hypothetical computer to aid the calculations which would be required in response. In 1937, he published the paper *On Computable Numbers* which as much as any single event can be seen as the start of the modern computer age.

▶ THE TURING MACHINE

The text outlined in detail a design for what became known as the 'Turing Machine', a computer whose basis lay at the heart of subsequent digital computers. It featured all aspects fundamental to modern computers such as the ability to read, write and erase data, a memory to store data, a central processing unit and the concept of a programme craeted through a series of mathematical instructions. The device as described was never built, but it is arguable that it has effectively been mass produced in an ongoing and modified form since the 1950s.

Turing's On Computable Numbers *can be seen as the start of the modern computer age*

▸ CRACKING ENIGMA

During the Second World War Turing's abilities were diverted to solving the algorithms behind the German 'Enigma' code. This was a machine code used by all branches of the German armed forces, but in a particularly complex form by the navy. Only when Enigma was broken, and the Allies could finally track the movements of German submarines, was the Battle of the Atlantic finally won. Turing was vital in cracking this code (which also involved work on a primitive computer called 'Bombe'). In the process he saved the lives of thousands of Allied seamen who could now divert their vessels away from the U-boats, thanks to intercepted instructions.

▸ THEORY INTO PRACTICE

After the war Turing turned away from theoretical mathematics and put his skills to use in the beginnings of the computer industry. He accepted a post at the National Physical Laboratory and was involved in the construction of the 'Automated Computer Engine' (ACE), an early digital computer. Shortly afterwards, in 1948, he worked on the 'Manchester Automatic Digital Machine' (MADAM) at the University of Manchester, then the computer with the largest memory in the world. As well as being involved in the physical construction of the machines during this phase, he also applied his knowledge of mathematics to the development of early programming languages.

▸ THE TURING TEST

Turing had no doubt that computers would not only play an increasingly important role in the lives of subsequent generations, but was also convinced they would eventually reach a level of sophistication whereby they could 'think' as well as humans.

To measure the point in the future at which this had been achieved he devised a test outlined in his 1950 paper *Computing Machinery and Intelligence*. This suggested what became known as the 'Turing Test', whereby a remote computer operator would have to ask questions of both a human and the 'intelligent' computer. If the operator could not distinguish the replies between the living person and the machine then the computer would have passed the test. Turing believed this point would come by the year 2000; arguably a premature estimate but an easier one to envisage now than fifty years ago. The test continues to play a part in debates about Artificial Intelligence.

▸ AN UNHAPPY END

Sadly Turing did not live long enough to see his predictions take off, particularly after the spectacular change brought about through the growth in popularity of the transistor and, later, the silicon chip. He is thought to have committed suicide shortly after his conviction for a homosexual offence, still criminalised at the time.

A CHAMPION OF COMPUTER SCIENCE

Of the many defining events of the twentieth century that changed the way live or our understanding of it, the digital computer must be somewhere near the top of the list. Charles Babbage (1791–1871) had earlier devised primitive mechanical computer models with his Difference Engines, and although the ultimate development of the computer industry was the result of the efforts of many individual scientists, Alan Turing can be remembered as a man who achieved more than many others in its early stages.

JONAS SALK

1914–1995

CHRONOLOGY • **1939** Salk graduates with an MD from New York University • **1942** Joins the University of Michigan influenza immunisation project • **1947** Begins research into poliomyletis at the University of Pittsburgh • **1955** Salk's vaccine is declared safe and effective in the United States

By the mid-point of the nineteenth century advances in medical science had seen many diseases banished by vaccination or limited through effective treatment. One which had not been dealt with, however, and which seemed all the more cruel, was polio, because it predominantly left in its wake a trail of paralysed children. Worse still, its incidence was increasing and in the 1940s and early 1950s there were horrific epidemics, particularly in the United States. The person who could find a vaccine to conquer polio would instantly become a national hero. Jonas Salk became that American idol, although not without backlash from some of those he had beaten to the prize.

▶ A GROUNDING IN IMMUNOLOGY

Before his victory over polio, Salk had in fact been successful in other areas of immunology. From 1942 he had worked under Thomas Francis Junior (1900–69) at the University of Michigan as part of a team developing an influenza vaccination. This treatment was eventually administered to members of the armed forces during the war. From 1947, Salk moved to the

The person who could find a vaccine to conquer polio would become a national hero

University of Pittsburgh where he continued work on improving the influenza vaccine. At the same time his attention was drawn to the disease of polio and the desperate need for some kind of immunisation.

There had been earlier attempts to develop a vaccine for polio but this had ended unfortunately. In 1935, a preventative treatment had initially appeared triumphant, but when trials were extended to ten thousand test cases it had proved both unsuccessful and in some instances dangerous, inducing full-blown cases of the disease.

▶ AMALGAMATING TREATMENTS

With safety in mind, Salk began developing a 'killed-virus' vaccine. This involved taking live samples of the virus and then killing it by soaking it in formaldehyde. He still hoped that the dead strain, when injected, would force antibodies in the human immune system to increase resistance to the disease to the point where any future exposure to a live virus would be harmless. In preparing this treatment, Salk built on the existing findings of other scientists. These included the discovery that there were three strains of the polio virus. Thus, any vaccine would have to be able to combat all versions. Indeed, it could be argued that Salk brought together a series of findings previously discovered by others rather than generate any new concept. Given the amount of fame and public gratitude which followed his success, it was this factor which caused bitterness within the scientific community.

▶ A VACCINE FOR POLIO

Salk began testing his virus first on monkeys, then on a small number of humans in 1952. He found, as he had hoped, that extra antibodies were produced as a result of the injection without noticeable side effects or vaccine-induced versions of the disease occurring. Salk published his findings in 1953 and submitted them for mass trial the following year. His former instructor, Francis, organised this trial on nearly two million children. The following year he reported it to be a great success. Salk was revered as a saviour. In 1955, the vaccine was approved for general release by the US government and, despite initial fears of another 1935-style disaster (some children had been contaminated by a defective batch) the vaccine went on to combat the disease. Additional safety measures introduced in the preparation of the treatment meant millions more were immunised with no ill effects.

▶ A RIVAL CONTENDER

While many who refused to credit Salk were motivated by jealousy, perhaps one who had cause to feel some genuine bitterness was Albert Bruce Sabin (1906–93). He developed a live attenuated vaccine for polio shortly after Salk and despite his insistence that his version was more effective and easier to administer, it was initially ignored. Eventually, after being forced to conduct trials in Russia, his version – taken orally – was widely adopted in place of the Salk jab, thus proving to be more successful.

THE SALK INSTITUTE

The now celebrated American went on to become director of the Institute for Biological Studies in La Jolla, California. It was later reopened as the Salk Institute and became famous internationally as an esteemed research centre. Due to the shunning he received from much of the scientific community after his fame, however, Salk later commented, 'I couldn't possibly have become a member of this Institute if I hadn't founded it myself'.

ROSALIND FRANKLIN

1920–1958

CHRONOLOGY • **1951** Franklin begins work as assistant to John Randall at King's College, Cambridge, alongside Maurice Wilkins • **1952** Describes the basic helical structure of DNA molecule • **1953** Her work is used, uncredited, in Watson and Crick's Nobel Prize-winning paper • **1958** Dies in London of cancer aged 37

Few of the stories in the history of science are as fascinating, or controversial, as the race to unravel the structure of DNA. The key to understanding the formulation of the deoxyribonucleic acid which helps make up the genetic information-carrying chromosomes, was the key to understanding life itself. The team which eventually claimed that 'prize' were the Cambridge duo of James Dewey **Watson** (b. 1928) and Francis Harry Compton Crick (b. 1916), 'secretly' aided by leaked information from Maurice Hugh Frederick Wilkins (b. 1916) at the

University of London. The breakthrough data Wilkins had shown Watson were the results of work performed by his estranged colleague at the University of London, Rosalind Franklin, who arguably lost her slot in history as a result. Franklin had graduated in 1941 with a good degree in chemistry from Cambridge, and before she took up her new position at King's College in 1951, had already made contributions to the understanding of the structure of graphite and other carbon compounds, and had carried out studies on the absorption properties of coals for the British Coal Research Association.

Rosalind Franklin was an unsung heroine of the study of genetics

▸ A TOUGH PLACE FOR WOMEN

Britain in the early 1950s, however, when the DNA race was won and lost, was still a tough place for women to make their mark in the workplace. Although there had unquestionably been improvements over the preceding half-century in the battle for equality of the sexes, much of the 'old school' attitude still persevered. In such a potentially hostile environment for women, then, it is no wonder Franklin chose to work on her own. Equally, it was clear she and her colleague Wilkins, who were both working on the DNA question under the directorship of John Turton Randall (1905-84) at King's College, simply did not get on. Their lack of co-operation was in stark contrast to the teamwork of Watson and Crick at Cambridge which ultimately proved so successful.

▸ A LONELY PIONEER

So, on her own then, Franklin pursued the task of figuring out the structure of DNA. She quickly made some impressive strides. The link between DNA and its basis as the mechanism for passing on hereditary information had already been reasonably well established. Franklin, like Watson and Crick, built on that knowledge, as well as that from a series of other scientists around the world who had also offered findings. The next step to understanding how DNA so successfully shared its data, both parties independently decided, was to understand its structure; from that should come the answer. Here, in many ways, Franklin had the advantage. She was an expert in X-ray diffraction techniques, a method which had been used for taking pictures of atoms in crystals and which was just starting to be applied to biological molecules. Franklin began examining DNA, then, via this means. The results of her investigations brought two important findings. Firstly, she realised that the 'backbone' of the molecule was on the outside, which Watson and Crick had at first missed, and was vital in eventually understanding its structure. Moreover, by 1952, Franklin had taken the clearest pictures of the molecules to date, which provided evidence of a helical, or spiral, structure. Watson and Crick would ultimately articulate a 'double helix' construction.

Indeed, it was while Franklin was trying to piece together the implications of these advances that they found their way to the Cambridge camp. Franklin's colleague Wilkins had access to her pictures and showed them to Watson. Immediately, Watson noted the evidence for the helical build, the vital piece in the jigsaw he and Crick had been trying to put together, and soon afterwards they made the breakthrough announcement that they had unravelled the structure of DNA.

AN ACHIEVEMENT DENIED

Although Franklin was already a distinguished chemist before she went to work with Randall, Watson's desultory description of her in his book on the discovery of the double helix helped to ensure that her role in the discovery was ignored. Shortly, afterwards, she contracted and died of cancer aged just thirty-seven. Watson, Crick and, moreover, Wilkins went on to be awarded the Nobel Prize for physiology in 1962 for their discovery. Even if the prize giving committee had intended to honour Franklin too, they would have been unable to do so: the Nobel Prize cannot be awarded posthumously.

JAMES DEWEY WATSON

B. 1928

CHRONOLOGY • **1947** Watson graduates from the University of Chicago, aged just nineteen • **1953** With Francis Crick, Watson proposes the notion of the double helix molecular model for DNA • **1962** Awarded the Nobel Prize for Medicine with Francis Crick and Maurice Wilkins

One of the most important scientific discoveries of the 20th century was made all the more interesting because the story behind it contained the key ingredients for the best drama: a race against time for a prize which would change the world, winners and losers, personality clashes, prejudice, a hint of sabotage, and a lingering question of 'what if?'

▸ THE DOUBLE HELIX

In fact, the story was so good, in 1968 James Dewey Watson, one of the main players,

published it under the title *The Double Helix*. Rather than use the account to smooth over the passionate tensions and heat of the moment friction encountered at the time of its setting fifteen years earlier, he cranked up the drama another level with previously untold tales of ambition, conscious Nobel Prize pursuit, obstructive authority figures and spiteful side players. Perhaps the most unfair portrayal was that of Rosalind **Franklin**, relegated to a bit player when the crucial discovery was in fact hers. It just added to the legend, of course.

The actual science behind the story, albeit in a

The key to DNA, Watson realised, was its shape: the now famous double helix

slightly less dramatic way, is no less interesting. Early in 1953, the American Watson and his English colleague, Crick, announced that they had, quite literally, unravelled the secret of life. They had concluded that DNA, which was known to carry the hereditary information at the basis of all life, had a 'double helix' structure. Moreover, it was the detail of this construction which allowed it to pass on its secrets so successfully.

▶ THE KEY TO DNA

Key to DNA were the four bases adenine (A), cytosine (C), guanine (G) and thymine (T) which the scientist Erwin Chargaff (1905-2002) had earlier studied and measured. He had noted that C and G were always present in the same quantities, and A and T followed a similar pattern. Watson and Crick suspected this indicated some kind of mutual attraction between the respective bases, which meant they would only 'pair' with their appropriate partner within the backbone of a DNA molecule. They gradually fitted these ideas into a structure but still could not work out how the DNA molecule passed on its information so accurately. Then came the vital viewing by Watson of Franklin's X-ray diffraction photographs of DNA, secretly shown to him by Franklin's University of London colleague Maurice Wilkins (b. 1916). 'The instant I saw the picture my mouth fell open,' Watson later said. DNA was made in a helical structure, in fact as

they would soon work out, a double helix, and over the following months Watson and Crick would finally realise why this was so important. When required to share its information, the two strands could literally uncoil themselves into two halves. This would leave the 'rungs of the ladder' containing A, C, G and T exposed. Naturally, they would seek to 'complete' themselves again but as A would only link with T, C with G and vice versa, this meant the strands would combine selectively with other matter in the cell in order to form two perfect copies of its original self.

It was beautifully simple and the discovery made Watson and Crick world famous.

▶ THE BASE FOR GENETICS

Watson, Crick and, perhaps a little contentiously, Wilkins went on to collect the Nobel Prize for physiology for their discovery in 1962. By that point Franklin was already dead. But the controversy surrounding the tale, thanks especially to Watson's later book, was far from finished. The discovery of the double helix was the starting point from which scientific exploitation of DNA and genetic information would grow rapidly in the latter half of the twentieth century. Controversies and benefits surrounding genetically modified food, ethical dilemmas concerning cloning, and court cases hinging on DNA evidence would be some of the later developments dependent on this breakthrough.

THE LEGACY OF WATSON

The discovery of the double helix was the starting point from which scientific exploitation of DNA and genetic information would grow in the latter half of the twentieth century. Controversies surrounding and the possible benefits of genetically modified food, ethical dilemmas concerning cloning, and court cases

hinging on DNA evidence would be some of the multitude of issues later spawned by this radical breakthrough.

In particular, Crick, Watson's colleague, would go on to make many additional contributions to the furthering of knowledge in the field.

STEPHEN HAWKING

B. 1942

CHRONOLOGY • **1960s** Hawking contracts the muscle-wasting motor neurone disease • **1971** Proposes the existence of mini black holes • **1974** elected fellow of the Royal Society, one of the youngest ever • **1977** Appointed Professor of Gravitational Physics at Cambridge • **1979** Appointed Lucasian Professor of Mathematics at Cambridge, a post once held by Isaac Newton

Stephen Hawking is one of the most notable, and certainly one of the most famous, theoretical physicians of the last fifty years. Acknowledged for his efforts in attempting to extend the General Theory of Relativity of **Einstein** (1879-1955), Hawking has sought to offer explanations in all areas of cosmology, particular the nature and properties of black holes.

▸ BIG BANGS AND CRUNCHES

After completing an undergraduate degree in mathematics and physics in 1962 and his PhD at Cambridge in 1966, Hawking worked with Roger Penrose (b. 1931) on black hole theory and the origins of the universe. The outcome of their analysis of Einstein's General Theory of Relativity was that the implied 'big-bang' start to the universe should have begun with a gravitational 'singularity,' where matter was infinitely dense and space-time infinitely curved. Equally, it should end in singularities called black holes, or even a 'big-crunch' where the universe contracts back to its original state.

Hawking sought to unite quantum theory with gravitational theory, an ambitious undertaking

▶ A 'RELATIVE' PROBLEM

The problem with these findings was that the existing General Theory of Relativity could not 'cope' with these singularities. So, Hawking sought to extend it by uniting quantum theory as applied to atomic-sized structures with the gravitational theory applied to the 'wider' universe (as outlined in Einstein's General Theory). The need for this combination was highlighted even further when Hawking suggested in 1971 the idea of the formulation of mini black holes immediately after the big-bang. These phenomena would have weighed up to a billion tons, and thus being susceptible to gravitational law, but only have been the size of a proton, thus having to obey quantum laws. This attempt to combine the two greatest physical theories has proved difficult, but did lead Hawking on to new developments in black hole theory.

In 1974 Hawking suggested that in line with his application of quantum theory, black holes, out of which it was previously believed nothing could escape, including light, and whose properties could never be known, could not actually be 'black' at all. Instead, they must effectively emit some energy where pairs of particles are separated, with negatively charged particles getting sucked into the hole and positively charged ones escaping as energy. This allowed the laws of thermodynamics also to be applied, uniting to some extent the quantum and classical principles. Eventually, the black hole would radiate away all of its energy and vanish.

Further implications of Hawking's 'quantum gravity' suggested, indeed, there might be no singularities at all so the known laws of physics would therefore always apply, and always have applied. This would also imply no beginning or end to the universe, or any boundaries to it at all.

Hawking's achievements in advancing our understanding of black holes and extending debate on the scientific origins of the universe are all the more remarkable because he has continued work in spite of being diagnosed with the debilitating motor neurone disease since his student days. This has left him in a wheel chair and unable to talk, meaning he now communicates through a computer. In spite of only being able to convey a maximum fifteen words per minute via this method he has written and published numerous articles and books on his subject.

Hawking is almost certainly as famous for his ability to convey complex scientific ideas on the origins and physics of the universe to the general public as he is for his original scientific thought. He has achieved the rare combination of being talked of in academia in the same breath as some of the all-time greats of physics, whilst having many of his ideas, to some extent at least, understood by those outside his immediate field due to his popular presentation of them. Most notable amongst his books written in this vein is the 1988 bestseller *A Brief History of Time: From the Big Bang to Black Holes*. Today, he continues to remain undaunted by such grand subjects, as demonstrated by the title of his 2002 publication, *The Theory of Everything: The Origin and Fate of the Universe*.

THE WORLD OF STEPHEN HAWKING

The combination of his writing for a popular audience, his radical ideas and his success over disability have made Hawking world famous. The scientist himself puts this down to people being, 'fascinated by the contrast between my very limited physical powers, and the vast nature of the universe I deal with.'

TIM BERNERS-LEE

B. 1955

CHRONOLOGY

- **1976** Berners-Lee graduates from Queen's College, Oxford
- **1978** Leaves Plessey to join D.G. Nash Ltd, Ferndown, England
- **1984** Takes up a fellowship at CERN, the European Particle Physics Laboratory, in Geneva
- **1989** Proposes a global hypertext project, to be known as the World Wide Web, designed to allow people to work together by combining their knowledge in a web of hypertext documents

Step back two and a half thousand years to the beginning of this book and consider for a moment the mysterious existence **Anaximander** (c. 611–547 BC) was grappling with. Now fast-forward to the present day and attempt to comprehend how science has changed our understanding of the world and our successful manipulation of the elements contained within it. Tim Berners-Lee, the English inventor of the World Wide Web, is the last entry in a long line of those who have built on the work of others to change a facet of our lives, resulting in a modern existence which would be every bit as incomprehensible to Anaximander.

▶ ON THE SHOULDERS OF GIANTS

Little has changed our relationship with the world as much as the science of computers in which Berners-Lee works, and the World Wide Web has become a key tool in that revolution. Unlike equally significant recent developments, such as the joining together of a network of computers, a network of networks in fact, into an

Few applications of science can have changed our world more than the computer

'Internet,' and the evolution of applications such as email which make use of this tool, the invention of the World Wide Web is particularly notable because it can be pinpointed to a sole creator, Berners-Lee. Rarely has the work of a single person had such a remarkable impact on business, research and individual lives as Berners-Lee's 1989 creation.

▸ THE INTERNET AND THE WEB

The World Wide Web is quite distinct from the Internet. The latter is the physical infrastructure through which data can be transmitted. The 'Web,' however, was the first means by which the world at large gained access to and the ability to share information across this Internet.

The idea first occurred to Berners-Lee as an offshoot of a 1980 program he wrote called 'Enquire.' The concept was simple: he wanted to keep track of electronic information by 'linking' words in certain documents with other documents on his computer. Thus, Berners-Lee could jump from one related document, or piece of information, to the next, with minimum effort. Over the following years, the Englishman began considering ideas for allowing him to link to documents on other people's computers, and from theirs to his, without the need for a central database. It was as a logical extension of this vision that in 1989 he proposed a project to be called the World Wide Web.

Berners-Lee wrote a simple common language called Hypertext Mark-Up Language (HTML) through which authors could prepare documents in a common format with the necessary links, a method for linking these pages across the Internet (Hypertext Transfer Protocol or HTTP) and an addressing system for identifying and accessing the pages via a Universal Resource Locator (URL). His next step was then to create a straightforward Graphical User Interface (GUI) via which ordinary, non-technical people could read and share in these pages. It was launched onto the Internet at large in 1991, and in no time people were linking their pages across the world in a completely 'uncontrolled' Web.

Berners-Lee, like the rest of the world riding on the back of his creation, has come a long way since graduating from the University of Oxford in 1976. His early career in Dorset gave little insight into the revolution which would follow. He then went on to work in Geneva as a software consultant at the European laboratory for particle physics called CERN, where he had his initial idea for Enquire. Today, Berners-Lee lives and works in the United States at the Massachusetts Institute of Technology in its Computer Science Laboratory. He heads the World Wide Web Consortium, or 'W3C', which aims to 'lead the web to its full potential.'

Thus another application of science, the World Wide Web, brought about through the far-sighted vision of a single scientist, has again changed the world. How many more times will science continue to do so over the next two and a half thousand years? Our guess would probably be no better than Anaximander's.

EVERYDAY LIFE AND THE WEB

Ongoing improvement in 'browsers,' in particular, as well as other technology, have facilitated easier access to the Web to the point today where literally tens of millions of people make use of it every day, a number which is still growing. Now if we want to buy a car, research an essay, listen to the radio or find out a weather report, amongst thousands of other things, it can all be done on the Web in a way that less than two decades ago was impossible.

SCIENTISTS A–Z

PICTURE CREDITS

Cover image by M.Kulyk / ©Science Photo Library Ltd, London
Images pages 8–24, 30, 36–50, 54–58, 62, 68, 74–82, 88–98, 104, 108–114,
118, 122, 126–132, 136–146, 152–154, 160–164, 168, 176–186, 194, 202
©Mary Evans Picture Library

Arcturus Publishing Limited has made every reasonable effort to ensure
that all permissions information has been sought and achieved as required.
However there may be inadvertent and occasional errors in seeking permission
to reproduce individual photographs for which Arcturus Publishing Limited
apologizes.